SEA CUCUMBERS OF BRITISH COLUMBIA, SOUTHEAST ALASKA AND PUGET SOUND

T0164100

ROYAL BRITISH COLUMBIA MUSEUM
H A N D B O O K

Sea Cucumbers

OF BRITISH COLUMBIA, SOUTHEAST ALASKA AND PUGET SOUND

PHILIP LAMBERT

UBC PRESS / VANCOUVER

Published by UBC Press in collaboration with the Royal British Columbia Museum.

Canadian Cataloguing in Publication Data
Lambert, Philip, 1945-
    Sea Cucumbers of British Columbia, Southeast Alaska and Puget Sound
    (Royal British Columbia Museum Handbook, ISSN 1188-5114)

    Includes bibliographical references and index.
    Copublished by: Royal British Columbia Museum.
    ISBN 0-7748-0607-9

    1. Holothurians--British Columbia. 2. Holothurians--Alaska. 3. Holothurians--Washington (State)--Puget Sound. I. Royal British Columbia Museum. II. Title. III. Series.
QL384.H7L35 1997   593.9'6'091643   C97-910406-8

UBC Press   Tel: 604-822-3259
University of British Columbia   Fax: 1-800-668-0821
6344 Memorial Road   E-mail: orders@ubcpress.ubc.ca
Vancouver, BC
V6T 1Z2

Printed in Canada.

CONTENTS

PREFACE

Why would I want to study sea cucumbers, a group many would consider to be slimy, obnoxious creatures? More often than not sea cucumbers are the butt of jokes like: Knock knock! Who's there? Seek. Seek who? Seek-who-cumber! "Ha! Ha! Very funny!" I retort as I look up from my microscope to see who is casting aspersions. I could rationalize this work with arguments based on economics and biodiversity but the simple answer is that, like most scientists, it is the joy of discovery that fuels my interest. After writing this book, I realize we know even less than I thought. Having formerly written a handbook about sea stars, it seemed appropriate to write a handbook on another closely related group. The sea cucumbers proved to be more challenging than sea stars. Many identifications in the previous literature were incorrect or suspect; the classification of some species was inconsistent; original scientific descriptions were meagre and in obscure journals, and there appeared to be a few undescribed species as well. All in all, this made it difficult even for a marine biologist to make routine identifications with any certainty. For a group of conspicuous marine animals, it seemed unusual not to have a good identification guide available. This alone seemed to be enough reason to pursue this project. What we needed was a guide that would bring together all the scattered information about each species and help the reader to make accurate identifications.

In order to write this book some original research was necessary to clear up the taxonomy of several species. Obviously it would be difficult to write a popular account of sea cucumbers if the scientists did not even have them straight! This included redescribing several species, describing new species and revising the classification.

INTRODUCTION TO SEA CUCUMBERS

This book covers species that live on the continental shelf from Skagway in southeastern Alaska to Puget Sound, and from the intertidal zone to a depth of 200 metres, or the edge of the continental shelf. Most of these species can be collected at low tide or by scuba diving. A few, such as *Pentamera* and *Molpadia*, are common in soft sediments and are usually collected by dredge or trawl.

Many species in this guide can be identified by colour, shape and other external features, but some closely related species are difficult to tell apart and will require some dissections or microscope preparations. For that reason, I have illustrated skin ossicles and calcareous rings. Most of the photographs in this book are of live species in their natural habitat or in an aquarium, but some photographs are of preserved specimens.

Origins of Sea Cucumbers

The first unequivocal fossil evidence of sea cucumbers dates from the late Silurian period, about 400 million years ago. Unlike many animals, sea cucumbers have few hard parts, making their fossils difficult to find. Complete fossil specimens of only 12 species have been collected with all the hard and soft parts intact. Usually scientists only find fossil imprints of tiny calcareous, skin particles, not a whole animal. As a result, our knowledge of the evolutionary history of sea cucumbers is sketchy.

Sea cucumbers are part of the group Echinodermata, from the Greek meaning spiny (echino) and skin (derm). Echinoderms include sea stars, sea urchins, feather stars, brittle stars, sea cucumbers and a recent addition called sea daisies. Scientists discovered sea daisies in 1986 and placed this group in a new class of echinoderms, called Concentricycloidea. These tiny animals found in the deep sea off New Zealand (1000 m) and in the Caribbean (2000 m) have the characteristics of echinoderms. Whether these animals are sufficiently distinct to be placed in a new class remains controversial.

Although the classes of echinoderms are superficially different, they share some characteristics. All echinoderms have a hydraulic system (the technical term is water-vascular system) made up of tubes and valves that operate five rows of extendible tube feet. These five rows radiate from a central ring, an arrangement that biologists call pentaradial symmetry. The third characteristic of echinoderms is an internal skeleton made up of calcite, a crystalline form of chalk. The skeletons vary from the solid plates of a sea urchin to minute particles scattered through the skin of a sea cucumber.

These three characteristics — a water-vascular system, pentaradial symmetry and an internal calcareous skeleton — make echinoderms distinct in the animal kingdom. Yet the evolutionary history of this group remains controversial. There is still no consensus about the origin of echinoderms or even the evolution of the six classes within the phylum. Some scientists conclude that sea cucumbers and sea urchins arose from a common ancestor. Others, based on the form of the larva, think that brittle stars and sea urchins have more in common. Another study suggests that sea cucumbers are the most primitive of the echinoderm classes. Although there is no agreement yet on which classes are most closely related, some recent work on DNA supports the idea that sea cucumbers are most closely related to sea urchins.

Characteristics of Sea Cucumbers

Sea cucumbers, known technically as the Holothuroidea, are echinoderms with a soft body wall containing circular and longitudinal muscles, and a skeleton made up of isolated calcite particles. Typically, a sea cucumber is an elongate cylinder lying on its side with the mouth at one end and the anus at the other. Five rows of tube feet run the length of the body. Around the mouth there are one or two circles of feeding tentacles, which are actually modified tube feet. In the body wall, microscopic particles of calcite, called ossicles, represent a vestige of the normal echinoderm skeleton.

Figure 1 shows seven variations of the basic body plan, typified by the genera: *Cucumaria*, *Psolus*, *Parastichopus*, *Thyonidium*, *Pentamera*, *Molpadia* and *Chiridota*. For example, in *Chiridota* all that remains of the tube feet is a ring of feeding tentacles (modified tube feet) around the mouth.

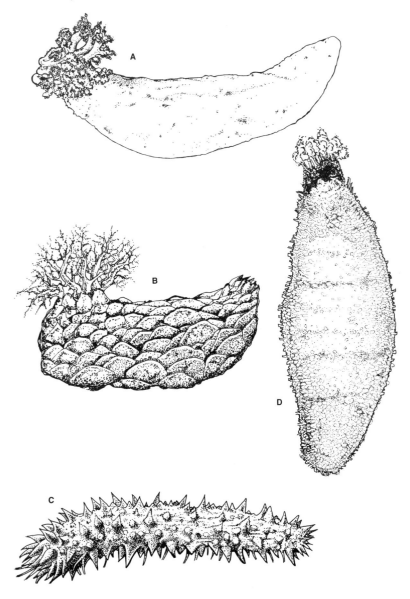

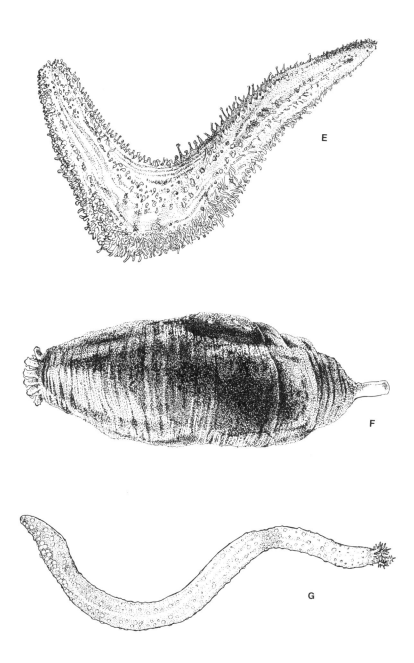

Fig. 1.Seven body forms of sea cucumbers: **A**) *Cucumaria* **B)** *Psolus*
C) *Parastichopus* **D)** *Thyonidium* **E)** *Pentamera* **F)** *Molpadia* **G)** *Chiridota*

External Anatomy

Tube feet, also known as podia, are extensions of the water-vascular system. Each podia usually consists of a cylindrical shaft with a sucker at the tip. Most lie in five rows extending the length of the body. In many species the three ventral rows are usually more robust than the two on the dorsal side. Like other parts of the body, the tube feet are quite variable. In some species, such as *Leptosynapta clarki*, tube feet are absent. On the dorsal side of *Parastichopus californicus* the pointed bumps, or papillae, are modified forms of tube feet.

The feeding tentacles, being part of the water-vascular system, can be extended by hydraulic pressure. The four basic types of tentacles are dendritic, peltate, digitate and pinnate (Figure 2). Dendritic tentacles gather small particles suspended in the water. Particles adhere to a coating of mucus on the tentacle, then the sea cucumber places it into its mouth and removes the food. This is suspension feeding. *Cucumaria miniata* is a common suspension feeder.

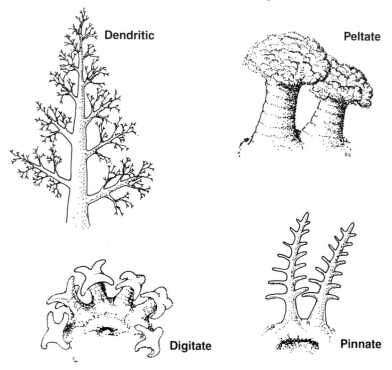

Dendritic   **Peltate**

Digitate   **Pinnate**

Fig. 2. Four basic types of tentacles: dendritic, peltate, digitate and pinnate.

Sea cucumbers that ingest sediment as they roam across the ocean floor, such as *Parastichopus californicus*, have moplike, peltate tentacles. The animal presses the tentacles onto the substratum and particles adhere to mucus. It then retracts the tentacle and inserts it into its mouth to remove the particles. The organic matter is digested from the sediment as it passes through the gut.

Sea cucumbers that ingest sediment as they burrow below the surface have digitate (e.g. *Molpadia*) or pinnate (e.g. *Leptosynapta*) tentacles. These short, fingerlike tentacles push sediment into the mouth

The introvert or "neck" region of those species with dendritic tentacles is a thin-walled area just behind the tentacles. The introvert and tentacles are pulled into the body cavity by the contraction of five retractor muscles, thus preventing predatory fish from nipping the tentacles.

Internal Anatomy

Figure 3 shows the internal anatomy of one type of sea cucumber in the genus *Cucumaria*. To dissect a preserved sea cucumber, determine the dorsal side, which is usually slightly darker or with tube feet less well-developed, and make a lengthwise incision. Use sharp-pointed scissors and make the cut to the left of the midline. This will leave the dorsal mesentery intact. The digestive tract in this genus is two to three times the length of the body. This length increases the area for absorption of nutrients. The stomach is a slightly expanded part of the gut. The anterior part of the intestine hangs from the mid-dorsal body wall by a transparent sheet of tissue, or mesentery. Other mesenteries connect the next part of the intestine to the left side and finally to the ventral body wall. The position of these mesenteries varies in different groups of sea cucumbers and this arrangement is an important character in classification.

Five bands of longitudinal muscle — and in some groups, the tentacle retractor muscles — are plainly visible. The rest of the body wall consists of circular muscles, connective tissue and skin. The action of these circular and longitudinal muscle layers produces a wormlike or peristaltic action. Imagine a long, thin, water-filled

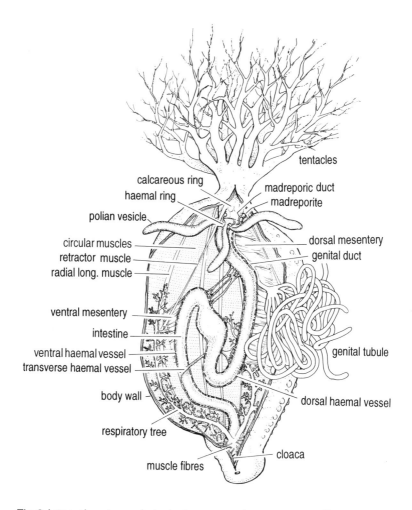

Fig. 3. Internal anatomy of a typical sea cucumber, represented by *Cucumaria*. After Pawson (1970)

balloon. When you squeeze one end the other extends forward. Some species use these waves of contraction to move along the ocean floor or through the mud. Others walk on their tube feet. Respiratory trees are the "lungs" of a sea cucumber. These hollow, branched organs lie inside the body cavity on either side of the intestine. The base of the tree connects to a muscular cavity, or cloaca (Figure 3). Circular muscles, or sphincters, close each end of the cloaca. A sea cucumber "breathes" by expanding the cloaca to draw oxygenated water in through the anus. The posterior sphincter closes then the cloacal muscles contract to force water up into the respiratory trees. Oxygen is transferred across the thin membrane into the fluids of the body cavity. When the oxygen is depleted, the main body wall contracts to squeeze water out of the trees. Although not occurring in our local species, some tropical forms have clusters of tubules, called Cuvierian tubules, attached to the base of the respiratory tree. When disturbed, the animal expels these sticky tubules, which appear to act as a repellent.

Circling the oesophagus just behind the mouth is the ring canal. This part of the water-vascular system may have sac-like extensions called polian vesicles. They are thought to function as expansion chambers for the water-vascular system. The ring canal has one or more stone canals attached to it. These are calcified tubes with a perforated swelling at the end called the madreporite. The function of stone canals is not clear, but it may have something to do with maintaining the fluid levels in the water-vascular system. The number and position of stone canals is another useful taxonomic character.

Five radial canals emerge forward of the ring canal sending branches to the tentacles, then curve back to run the length of the body between the longitudinal and the circular muscle layers. Short branches go to the tube feet and a small sac, or ampulla, marks the base of each tube foot. When contracted, these ampullae force fluid into the tube foot, causing it to extend. In some species like *Parastichopus californicus*, the base of each tentacle also has an ampulla. In the subclass Apodacea, tube feet are entirely lacking.

The simple nervous system consists of a nerve ring next to the ring canal, radial nerves and their branches. There is no central brain, but ganglia occur at various places on the main nerve branches. A nerve net in the skin detects touch and chemical stimuli and some species have sensory buds on the skin. Some species such as *Molpadia intermedia* have a balance organ, called a statocyst, that detects whether the animal is upright or not. Besides these examples there are few other discrete sense organs in sea cucumbers.

The blood system has no heart to move the fluid through the vessels, so it cannot really be called a circulatory system. The main blood channels form a network, called the rete mirable, running beside the intestine and respiratory trees. Most of the vessels are thin-walled spaces rather than discrete vessels. Blood cells also occur in the fluid of the body cavity: the coelom. Contraction of the body wall muscles pushes the coelomic fluid back and forth in a sluggish type of circulation. Oxygen from the respiratory trees diffuses directly into the coelomic fluid which also picks up nutrients directly from the intestine.

Sea cucumbers have haemoglobin, a unique feature among echinoderms. Haemoglobin is the chemical compound in blood cells that picks up oxygen from the environment and releases it to the body cells. Vertebrates also use haemoglobin. The blood cells of some lower vertebrates, like hagfish and trout, are remarkably similar to those of some sea cucumbers. Although it is tempting to conclude that these similarities imply a close evolutionary link between echinoderms and lower vertebrates, it is more likely to be the result of parallel evolution.

The calcareous ring is one of the few hard parts of a sea cucumber. It is comprised of a series of plates, usually 10, joined side by side like a collar around the oesophagus. The tentacle retractor muscles attach to this apparatus. The plates vary in shape in different species — some plates have long tails and others have anterior projections (Figure 4). The shape of the ring is important in the classification of sea cucumbers. For example, all those that have long posterior tails on the ring are placed in the same family. Each plate may be a solid piece, or in some species, a mosaic of smaller segments. Being one of the few hard structures in a sea cucumber, the calcareous ring is often the only part that fossilizes, thus providing a way of relating extinct and living forms.

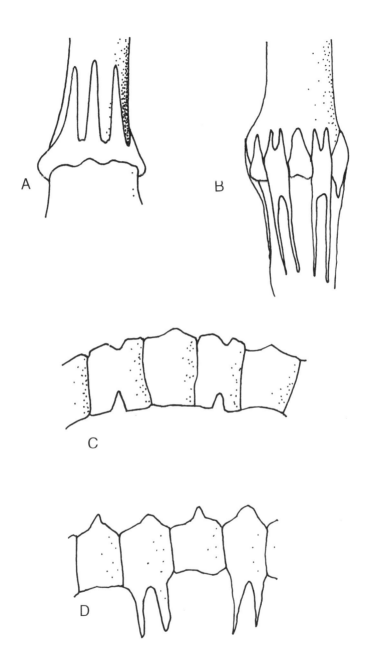

Fig. 4. Calcareous ring types: **A)** Family Cucumariidae
B) Family Phyllophoridae **C)** Family Stichopodidae **D)**Family Caudinidae

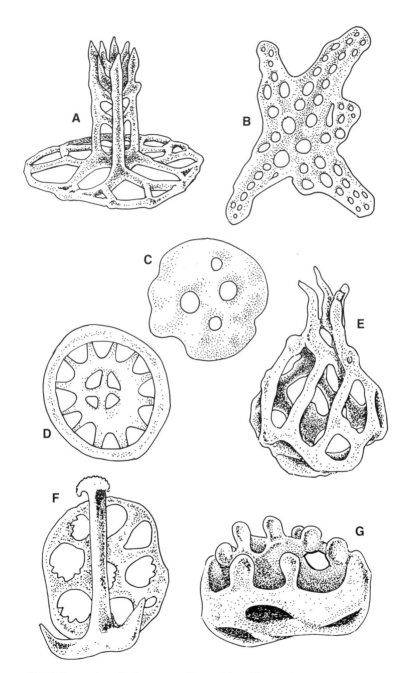

Fig. 5. Examples of calcareous skin ossicles: **A)** table from *Parastichopus*
B) perforated plate from *Pentamera* **C)** button from *Pseudocnus*
D) wheel from *Chiridota* **E)** basket from *Eupentacta* **F)** anchor and plate
from *Leptosynapta* **G)** cup from *Eupentacta*

The rest of the sea cucumber skeleton is represented by isolated pieces of calcite, embedded in the outer layers of skin. These microscopic ossicles, along with the calcareous ring, are extremely important in verifying the identity of a species. Ossicles are complex and varied in shape (Figure 5); they are quite beautiful. A description or an illustration of them is an important part of a scientific account.

Although ossicles are microscopic, one can often feel them protruding through the sea cucumber's skin. What is the function of these ossicles? Ossicles may protect the tiny juvenile since they are relatively large and armourlike in an animal only a millimetre long. They probably provide only a minimal degree of protection in the adult animal. In burrowing forms, like *Leptosynapta*, ossicles shaped like anchors protrude through the skin and provide traction as the animal crawls through the mud.

Special cells called sclerocytes (from the Greek *scleros,* meaning hard) secrete ossicles. The cell first creates a simple rod from which branches extend in various directions. The resulting shapes are characteristic for each species. For example, *Parastichopus leukothele* has an ossicle shaped like an inverted table (Figure 6).

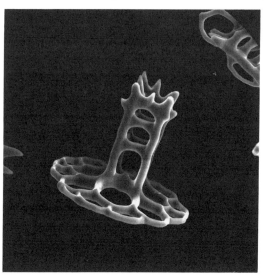

Fig. 6. Table ossicle from the skin of *Parastichopus leukothele*, taken with a scanning electron microscope. Diameter of disk about 100 μm.

Pseudocnus curatus has smaller, buttonlike ossicles (Figure 7).

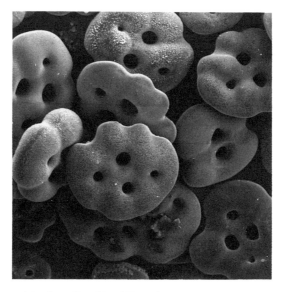

Fig. 7. Button ossicles from the skin of *Pseudocnus curatus,* taken with a scanning electron microscope.

Taxonomists use these ossicle shapes as one way of telling the species apart. Closely related species that have recently evolved from a common ancestor have very similar ossicles. The subtle differences between them can be difficult to detect. Ossicle shapes also vary within a single animal. Those in the dorsal skin can be larger on average than those in the ventral skin. Tube foot and tentacle ossicles can also be quite different. Furthermore, as the animal matures, the ossicles change from a complex juvenile form to a simpler adult form. Even experts sometimes find it difficult to separate closely related species.

Fortunately, recent developments in DNA research are making it easier to identify closely related species when traditional characters have proven inconclusive. We are now able to isolate an equivalent segment of the DNA molecule from several specimens and determine the exact sequence of bases. If the difference in the sequences is more than 2%, it is usually a good indication that they are different species. Some of the taxonomic changes in this book are based partly on DNA evidence.

You should be able to identify most of the species in this book from external characters; however, to verify some of the more obscure ones, it may be necessary to preserve some specimens for later analysis. For a specimen to be most useful it should first be relaxed or anaesthetized. This reduces the contraction of the specimen when fixed, and ideally leaves the tentacles extended. I usually use propylene phenoxetol, which can be ordered from some scientific supply houses. Prepare a stock solution of 0.015% in tap water (15 ml in 1 litre) and dilute with nine parts cold sea water just prior to use. Other chemicals that can be used as an anaesthetic include a solution of 75 gm of magnesium chloride (MgCl2) per litre of tap water mixed 1:1 with sea water, or a sprinkle of magnesium sulphate (MgSO4) or menthol crystals directly onto the surface of sea water. To preserve specimens for future examination, you will need a fixative such as formalin or ethanol. I use a solution of 10% formalin in sea water that has been neutralized with sodium borate (borax). You can check the PH level with a piece of litmus paper. It is important to neutralize the solution because formalin is acidic and over time will dissolve the ossicles and other calcareous parts of the animal that are important diagnostic characters.

To identify a few of the species in this book, you will need to look at the ossicles. It is a relatively simple procedure to isolate them. You will need some household bleach, a glass slide, a coverslip, a small dropper or pipette, some mounting medium and a microscope. Cut a small piece of skin (about 2mm square) from the mid dorsal side between the rows of tube feet. On the slide, cover the skin sample with a drop of bleach. The flesh will dissolve away in about five minutes leaving the ossicles behind. Placing a fine-tipped pipette at the edge of the drop, carefully draw off the bleach. An alternative method is to let the bleach and ossicles dry on the slide. The ossicles will then stick to the slide.

For a good permanent preparation it is crucial to wash the ossicles with water at least three times to remove bleach crystals. This part is tricky because it is easy to suck up the ossicles while removing the water. I usually swirl the slide quickly in tight circles to cause the ossicles to collect in the centre making it easier to draw the water off. Using the alternate method of drying, the sample should be flooded with water several times until no bleach crystals are left behind when the slide is dried. If you only want a temporary wet mount, place a coverslip on the water-covered ossicles and view the ossicles under the microscope. For a permanent slide, remove the last wash-water and let it air dry or, if available, use a slide warmer to speed up the process. Once the water has evaporated, place two small drops of mounting medium (Canada Balsam or any synthetic substitute) on the sample and apply a coverslip. Once the medium has set, you have a permanent reference slide.

Reproduction

The reproductive organs of a sea cucumber consist of one or two tufts of elongated tubules in the forepart of the body cavity. They combine into a single duct leading to an external gonopore near the tentacles. Most sea cucumbers have separate sexes, but the sex is difficult to determine by examining external features.

Sea cucumbers usually spawn annually, either by broadcasting eggs or sperm into the surrounding water, or by brooding the fertilized eggs. Some environmental cue causes many individuals to spawn simultaneously, thus increasing the chance for successful fertilization. The spawning cue may be a certain temperature, a number of sunny days in a row, or the presence of a plankton bloom. A female *Cucumaria miniata* releases a cluster of green eggs as a buoyant pellet. Males release sperm nearby and fertilization takes place by chance in mid-water. The pellet of eggs eventually breaks apart and each egg develops into a swimming larva. The larva develops and grows for days or weeks in the plankton and then settles on rocky areas inhabited by the adults.

A few species retain their eggs, rather than releasing them. As the eggs emerge, the female collects and holds them underneath her body. Sperm shed by nearby males fertilize the eggs. The larvae develop for a few months until the juveniles are large enough to crawl away. *Cucumaria pseudocurata, C. vegae, Pseudocnus lubricus* and *P. curatus* all reproduce this way during the winter. *Leptosynapta clarki* broods its young internally.

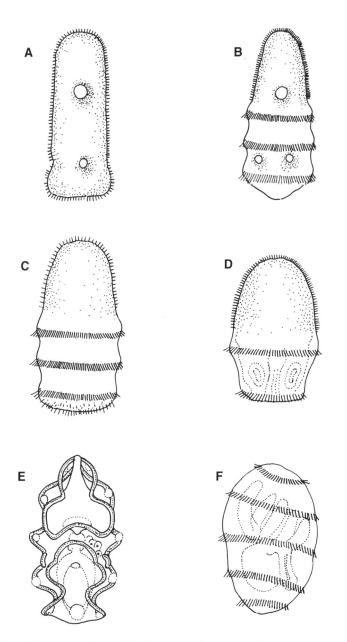

Fig. 8. Stages in the larval development of sea cucumbers:
A) four-day-old vitellaria of *Eupentacta quinquesemita*   **B)** vitellaria larva of
Pentamera populifera   **C)** vitellaria larva of *Psolus chitonoides*
D) vitellaria larva of *Molpadia* at 77 hrs   **E)** late auricularia stage of
Parastichopus californicus at 45-50 days   **F)** pelagic juvenile of
Parastichopus californicus at 55-60 days. After McEuen (1987). See © page.

Eggs range in size from a diameter of 200 μm to large yolky eggs of 1000 μm. Small eggs usually develop into planktotrophic larvae that feed on other plankton. Large eggs develop into lecithotrophic larvae that do not feed, but live off the yolk stored in the egg. Most sea cucumbers have this latter type of larva.

Sea cucumber larvae develop in one of two ways: indirectly or directly. Indirect developers undergo a radical metamorphosis or reorganization of the internal organs during their larval growth. Indirect developers first pass through an auricularia stage (Figure 8E); which has a series of looped ciliated bands for gathering food while it floats in the plankton. The organs of the auricularia undergo a radical rearrangement as the larva metamorphoses into a pelagic juvenile. This stage was formerly called a doliolaria. (Figure 8F.) Soon after this, the juvenile settles to the bottom.

Direct developers do not undergo an obvious, radical metamorphosis. Changes may take place slowly or some steps are greatly compressed in time. The early stage of a direct developer is a vitellaria larva (Figures 8A-D) which has enough stored yolk to sustain it without feeding. It develops into a pelagic juvenile with the basic adult body plan prior to settling onto the bottom. In brooding species, the larvae pass through the same stages before crawling away from the mother as a juvenile.

Predators

Sea stars and fish are the main predators of sea cucumbers. Some of the more mobile species, such as *Parastichopus californicus*, will rear up and roll away from a predator. When excessively disturbed it may even expel some of its internal organs out through the cloaca in an attempt to divert the predators attention. This is called evisceration, and it will be described under the Behaviour section. Other deep-sea cucumbers have movable flaps they use to swim off the sea floor for extended periods. In this way they escape bottom-dwelling predators and also feed on suspended particles in mid water. Some tropical species eject sticky tubules called Cuvierian organs that repel predators. For the most part, sea cucumbers have few defences except for hiding in crevices and under rocks. Some that occur in great abundance, like *Pseudocnus curatus*, use the strategy of safety in numbers. A predator will eat some individuals, but not all — like schooling fish, an individual plays the odds that its neighbour will get eaten, not it.

A few local species, such as *Cucumaria piperata*, contain a toxic chemical that deters some predators. The Kelp Greenling spits out pieces *Cucumaria piperata*; yet sea stars, on the other hand, seem oblivious to such poisons! Chemical defences are more commonly used by tropical species of sea cucumbers.

Physiology

Organic nutrients can be absorbed via the skin as well as through the gut. In all echinoderms, substances like amino acids — the building blocks of proteins — are actively transported through the skin. In one sea urchin studied, 13% of its food requirements was obtained by this method.

A sea cucumber respires through its tube feet, body wall and respiratory trees; the role of each tissue varies from group to group. In the Apodida (sea cucumbers without tube feet), most respiration is through the body wall, especially in those species lacking respiratory trees. *Molpadia intermedia* has a thicker, less permeable skin, but it has well-developed respiratory trees.

Behaviour

Suspension-feeding sea cucumbers seldom move. They use suckered tube feet to anchor themselves under rocks or onto hard surfaces, and creeping movements to adjust their position. Surface grazers, like *Parastichopus californicus*, must keep moving to reach new food sources. *P. californicus* propels itself forward slowly by creating waves of contraction along its body. Some species arch forward and others use their tentacles to anchor the anterior end while the posterior is drawn up. Others, like *Paracaudina*, burrow by contracting the circular muscles at the anterior end causing a sudden thrust of the "nose", like a small battering ram.

The process by which a sea cucumber throws out its guts is called evisceration. The reasons for this bizarre behaviour are not always clear. There is no doubt that certain species expel their internal organs when handled, or when subjected to abnormal temperatures in aquaria. Evisceration occurs in one of two ways: some species expel their guts through the anus, others through a slit in the anterior end of the body wall.

There is disagreement among researchers about whether or not evisceration occurs naturally. There is evidence that some sea cucumbers undergo a type of dormancy in the autumn, characterized by a cessation of feeding and locomotion, and an apparent loss of internal organs. Swan (1961) reported that 49 of 81 specimens of *Parastichopus californicus* had incomplete visceral organs in the fall. He concluded from this evidence that these specimens had spontaneously eviscerated. Fankboner and Cameron (1985) re-examined this phenomenon, and concluded that rather than expelling the visceral organs, the sea cucumber resorbed them. The autumn dormancy correlates with shorter days and a drop in plankton production — but a direct cause has not been established.

In the fall, a high proportion of *Eupentacta quinquesemita* individuals have no guts, or only a partially developed gut (Byrne 1985a). In this case, divers observed expelled guts in the habitat, in greater numbers than at other times. This species expels the gut rather than absorbing it. Some researchers suggest that evisceration may be a mechanism for getting rid of internal parasites, but this may not be the whole story. Eviscerated organs may be a way of diverting the attention of a predatory sea star, or a way of excreting waste materials that have accumulated in the gut. It may also play a role in dormancy by reducing metabolic costs during adverse conditions. The cucumber could save energy by not having to maintain a large set of organs.

The body wall of a sea cucumber contains layers of muscle and microscopic calcareous ossicles. Ossicles are embedded in a peculiar type of fibrous tissue unique to echinoderms. This "catch connective tissue", can be made stiff or soft by nervous control in a matter of seconds.

If you have ever handled a *Parastichopus californicus*, you will know that it can be stiff and plump one minute, only to transform a few minutes later to a soft, droopy bag of liquid. This is catch connective tissue in action. The soft phase facilitates normal movement. The firm phase is activated to support feeding tentacles or to anchor the animal in a crevice when attacked by a predator. It is much more energy efficient to use this tissue to maintain a feeding posture than to use prolonged muscle contraction. Scientists suggest that the success of the echinoderms may be related to the evolution of this tissue.

Parasites and Commensals
Parasitic forms of flatworms and snails live inside some sea cucumbers, and can damage some the sea cucumber's internal organs. Flatworms inhabit the fluids of the body cavity and upper intestine. Shell-less snails that look like worms attach to the outside of the intestine. Other shelled snails attach to the outside of the animal and suck the animal's body fluids.

Sea cucumbers are also accompanied by commensal organisms: animals that are closely associated with a host, but do not harm it. Scale worms that mimic the colour of a sea cucumber, crawl on the cucumber's skin; and small clams attach to the outside of the cucumber's body wall.

The known parasites and commensals of each species are described in the Biology Sections of each account. Many more probably exist, but have yet to be documented.

Economic Importance
Parastichopus californicus is the only species of sea cucumber harvested in British Columbia. The body wall of *Parastichopus californicus* is thin compared to the preferred Japanese sea cucumber, *Stichopus japonicus*; however, there is still demand for this west coast species. The five muscle strips and the body wall are utilized. Harvesting is by scuba because most *Parastichopus californicus* inhabit rocky areas where dredges cannot be used successfully.

Starting in 1980, harvesting in British Columbia was restricted to a few south coastal areas. The tonnage of landings appeared to peak in 1987-88; however, the statistics were initially based on the weight of the whole animal and has since changed to split weight. In 1986 a quota system was introduced, beginning with a coast-wide quota of 1500 tonnes from 1986 to 1988. This was decreased to 800 tonnes for the four years between 1989 and 1992, then down to 650 tonnes in 1993 and to 575 tonnes in 1994. Final data for 1994 indicate a total catch of 196 tonnes with a landed value of $976,000. In 1994 the average price paid for sea cucumber was $4.97 per kilogram.

Five harvesting areas were set up by the Department of Fisheries and Oceans (DFO). In 1994 the Pacific Sea Cucumber Harvesters Association (PSCHA) proposed a voluntary Individual Quota (IQ) fishery which would allow for a more controlled, safer harvest and a better product than the "shotgun" approach. DFO introduced a two-year pilot program of an IQ fishery in September of 1995. The IQ plan reduces the competition to be the first to harvest the resource in an area. Fishers can also take the time to process the catch properly and get it to market in good shape with a resulting higher price. DFO will also have a closer control over the total quota and will not be faced with accidental overfishing. The estimation of total sustainable catch is not a precise science, so the quotas are somewhat arbitrary. Each area is opened once every three years to allow for stocks to recover; however, the growth rate and age to maturity are still poorly understood.

Classification

Classification in this book is based on Pawson (1982). Biological classification is a hierarchical system just like that used in a library where all the books on a certain subject are in the same section. The classification system was introduced by Carl Linnaeus, a Swedish botanist, in 1758. Organisms with similar characteristics are grouped together in classes, orders, families and genera.

A species name is made up of two Latinized words, e.g. *Cucumaria miniata*. The first word is the genus, the second is the species. The genus is always upper/lower case, and the species is always lower case. Optionally, the name of the describer and the year the paper was published follow the species name. Parentheses around the author indicate that the species has been moved to another genus since the original description. In these cases I have indicated the original author and also the author of the revision.

A species may be defined as a set of similar individuals capable of breeding successfully with one another in nature and reproductively isolated from other sets. This is one of the accepted definitions, but it is an area of constant debate that cannot be adequately covered here. Traditionally, physical characters have been used to define species, but with the introduction of new genetic techniques scientists are discovering so-called cryptic species that externally appear identical. We should expect the discovery of many new species in the next few years using these techniques.

I have used only a few common names, mostly because few exist, but also because scientific names are more exact. I originally attempted to make up common names for all the species but as one reviewer pointed out many of them were longer than the scientific name and not very distinctive, especially for species that are variable in colour. Even in the scientific world, a single species may be described by two different authors, resulting in what are called taxonomic synonyms. In this case, the name that was published first is given priority and the newer name becomes what taxonomists call a "junior synonym". I have indicated some synonyms that may still be in the current literature and could cause confusion. To find out the complete history of name changes, however, one must consult more technical literature. When known, I have provided the etymology, or derivation of the species name.

Conservation

P. californicus is the only species of sea cucumber harvested commercially. Regulations are in place to try to ensure a sustainable harvest, but the effects of harvesting on wild populations are not well understood and should be monitored closely. Other non-commercial species of sea cucumber described in this book, like all marine organisms, are vulnerable to a variety of environmental conditions that affect the health of the marine ecosystem.

Each of us can contribute to the well-being of sea cucumbers and all other marine species by following some general ethical principles. As much as possible, observe but do not disturb animals in their natural habitat. If you must collect, keep specimens alive in an aquarium and return them to their natural habitat at the end of your study. Marine animals do not survive well in the trunk of a car and soon produce a rather unpleasant odour! Besides, they will never look as colourful and attractive as they do in the sea. Hopefully this book will allow you to identify the species without having to sacrifice any. On the beach, try to avoid trampling on any species and return rocks to their original position, while being careful to minimise damage to the animals under the rocks.

Unfortunately more than 50% of the world's human population inhabits the narrow strips of land next to the most productive part of the ocean — the coastal wetlands and the adjoining continental shelf. In the coastal environment of British Columbia and adjacent areas, habitat loss and pollution by human activity are the main threats to marine invertebrate diversity. As coastal habitats such as eel-grass beds and marshes are modified or filled in, the overall productivity of the marine environment can suffer. This has a ripple effect through the system that is not easy to measure or predict; consequently, the health of the system can erode gradually without being detected. Hopefully, the detailed information in this book will enable anyone to document the diversity of at least one small part of the marine biological diversity on our coast. Only by carrying out detailed baseline studies now, can we detect when changes in biodiversity occur in the future.

Species Covered

Following is a list of all the known species of sea cucumbers between Skagway, Alaska (58°45' N) and Puget Sound (47° N) and from the intertidal zone to the edge of the continental shelf at a depth of 200 metres.

CLASS HOLOTHUROIDEA
Subclass Aspidochirotacea
Order Aspidochirotida
Family Stichopodidae
Parastichopus californicus (Stimpson, 1857)
Parastichopus leukothele Lambert, 1986
Family Synallactidae
Pseudostichopus mollis Théel, 1886
Synallactes challengeri (Théel, 1886)
Subclass Dendrochirotacea
Order Dendrochirotida
Family Psolidae
Psolidium bidiscum Lambert, 1996
Psolus chitonoides Clark, 1901b
· *Psolus squamatus* (Koren, 1844)
Family Cucumariidae
Subfamily Cucumariinae
Cucumaria frondosa japonica (Gunnerus, 1767)
Cucumaria miniata (Brandt, 1835)
Cucumaria pallida Kirkendale and Lambert, 1995
Cucumaria piperata (Stimpson, 1864)
Cucumaria pseudocurata Deichmann, 1938b
Cucumaria vegae Théel, 1886
Pseudocnus curatus (Cowles, 1907)
Pseudocnus lubricus (H.L. Clark, 1901b)
Subfamily Thyonidiinae
Ekmania diomedeae (Ohshima, 1915)
Thyonidium kurilensis (Levin, 1984)
Family Phyllophoridae
Subfamily Thyoninae
Pentamera lissoplaca (Clark, 1924)
Pentamera populifera (Stimpson, 1864)
Pentamera pseudocalcigera Deichmann, 1938b
Pentamera trachyplaca (Clark, 1924)
Pentamera sp. A
Pentamera sp. B

Thyone benti Deichmann, 1937
Family Sclerodactylidae
    Eupentacta pseudoquinquesemita Deichmann, 1938b
    Eupentacta quinquesemita (Selenka, 1867)
Subclass Apodacea
Order Molpadiida
Family Molpadiidae
    Molpadia intermedia (Ludwig, 1894)
Family Caudinidae
    Paracaudina chilensis (Muller, 1850)
Order Apodida
Family Synaptidae
    Leptosynapta clarki Heding, 1928
    Leptosynapta transgressor Heding, 1928
Family Chiridotidae
    Chiridota albatrossii Edwards, 1907
    Chiridota discolor Eschscholtz, 1829
    Chiridota laevis (Fabricius, 1780)
    Chiridota nanaimensis Heding, 1928

KEY TO SHALLOW-WATER SPECIES OF HOLOTHUROIDEA FROM BRITISH COLUMBIA, SOUTHEAST ALASKA AND PUGET SOUND

Dissection or ossicle preparation may be necessary.

    To use the key, determine which one of the characteristics in #1 best fits your specimen. If your choice leads to the name of a species, then go to the figure indicated, for a full description. If your choice leads to a number, move to that number and once again pick the best description. If neither of the choices fit, return to previous set of characters (the number in parentheses) and see if another choice might be better. Repeat this process until you reach a species name then flip to its detailed description and check to see if it fits. Remember that a key is designed only as an aid to identification. You should always check the final choice with a detailed description.

1(0) U-shaped, or strongly curved, body even when relaxed (Figs. 1E, 46 and 48) 2

Long, tapering, grey body, with smooth wrinkled skin and no tube feet .... *Paracaudina chilensis* (Fig. 64)

Cucumberlike body (Figs. 1A and 1D) 3

Wormlike body (Fig. 1G) 16

Domed-shaped body with shinglelike scales, flat ventrally (Fig. 1B) 17

Body with papillae dorsally and tube feet on the ventral side (Fig. 1C) 19

Smooth sausage-shaped body with nipplelike tail (Fig. 1F) ....*Molpadia intermedia* (Fig. 62)

2(1) Tube feet conical in shape, skin ossicles are large, triangular perforated plates ....*Pentamera pseudocalcigera* (Fig. 48)

Tube feet straight sided; skin ossicles are round to star-shaped with a tall, narrow central spire and curved supporting tables with a tall spire ....*Pentamera populifera* (Fig. 46)

Tube feet straight sided; skin ossicles are round to triangular with a broad, low central spire and curved supporting tables with a low spire ....*Pentamera sp. A* (Fig. 52)

Tube feet straight sided; skin ossicles are small roundish tables with four main holes and four smaller holes; a two-pillared spire, and supporting table with a simple low spire ....*Pentamera sp. B* (Fig. 54)

3(1) Tube feet scattered all over body 4

Tube feet scattered dorsally, in rows ventrally 7

Tube feet in 5 series 8

4(3) Tentacles 10 or 15 5

Tentacles 20 or more 6

5(4) Tentacles 8 large, 2 small ....*Thyone benti* (Fig. 56)

Tentacles 10 large, 5 small ....
.............................*Ekmania diomedeae* (Fig. 40)

6(4) Tentacles peltate (Fig. 2) ...*Pseudostichopus mollis* (Fig. 13)
Tentacles dendritic (Fig. 2) ....
.........................*Thyonidium kurilensis* (Fig. 42)

7(3). Body black or grey; skin ossicles are mostly smooth 4-holed buttons (Fig. 37) ....*Pseudocnus curatus* (Fig. 36)
Body white to dark brown in colour; button ossicles have a knobbed surface, lobed margin, and 4 or more holes (Fig. 39)
.........................*Pseudocnus lubricus* (Fig. 38)

8(3) Tentacles equal in size 9
Tentacles 8 large, 2 small 11

9(8) Body orange; ossicles are slightly tapered plates (Fig. 27) .
............................*Cucumaria miniata* (Fig. 26)
Body black or grey; ossicles are large, irregular plates with spiny edges (Figs. 24 & 25) ....
..................*Cucumaria frondosa japonica* (Fig. 23)
Body pale orange or white, or white with black spots; ossicles are plates with serrated edges and tapered at one end to a spiny handle (Figs. 29 & 31) 10

10(9) White with black spots; one madreporic body ....
.........................*Cucumaria piperata* (Fig. 30)
White to pale orange, no spots; 5 or more madreporic bodies ....*Cucumaria pallida* (Fig. 28)

11(8) Calcareous ring with long posterior projections (Fig. 4C) ..12
Calcareous ring only with anterior processes (Fig. 4A) ..15

12(11). Skin ossicles mostly perforated plates (Figs. 45 & 49), or tables (Fig. 55) 13

Most of the skin ossicles are large complex bodies (Fig. 61)
....14

13(12) Ossicles are diamond-shaped plates ....
..........................*Pentamera lissoplaca* (Fig. 44)
Plates with complex knobbed surface on 1 side ....
.......................*Pentamera trachyplaca* (Fig. 50)
Small roundish tables with 4 main holes and 4 smaller holes
and a two-pillared spire ....*Pentamera sp. B* (Fig. 54)

14(12) Ossicles include baskets (Fig. 5E) ....
.....................*Eupentacta quinquesemita* (Fig. 60)
Ossicles include cups (Fig. 5G) ....
..............*Eupentacta pseudoquinquesemita* (Fig. 58)

15(11) Ossicles are flat oval plates (Fig. 33) ....
.....................*Cucumaria pseudocurata* (Fig. 32)
Narrow rod-shaped plates (Fig. 35) ....
.............................*Cucumaria vegae* (Fig. 34)

16(1) Ossicles are anchors and anchor plates (Fig. 5F) ....
.......................*Leptosynapta spp.* (Figs. 66 & 68)
Ossicles are wheels (Fig. 5D) ......*Chiridota spp.* (Fig. 70)

17(1) Small tube feet protruding through dorsal scales, in rows
ventrally ....*Psolidium bidiscum* (Fig. 17)
No tube feet among dorsal scales, ventral series in rows ..18

18(17) Tentacles and scales white ....
.............................*Psolus squamatus* (Fig. 21)
Tentacles and scales red or orange ....
.............................*Psolus chitonoides* (Fig. 19)

19(1) Body beige with purple tinge, dorsal papillae slender,
ossicles are cross-shaped tables with tall spire (Fig. 16),
tentacle ampullae absent ....
.......................*Synallactes challengeri* (Fig. 15)

Body mottled brown, red or orange, table ossicles with circular base and spire of 4 columns (Fig. 5A), tentacle ampullae present 20

20(19). Body mottled red or brown, rarely white; shallow subtidal .
.......................*Parastichopus californicus* (Fig. 9)
Body orange with white papillae, usually in water deeper than 30 m. ....*Parastichopus leukothele* (Fig. 11)

SPECIES DESCRIPTIONS

Family Stichopodidae

Parastichopus californicus (Stimpson, 1857)
Parastichopus leukothele Lambert, 1986

External Features: Body with tube feet in rows ventrally, papillae dorsally. Body wall soft and pliable. Tentacles, 20, equal, peltate.
Internal Features: Tentacle ampullae present. Retractor muscles absent. Respiratory trees Y-shaped. Rete mirable present. Posterior mesentery attached to right ventral body wall. Gonad double tuft. One madreporic body, attached to dorsal mesentery. Cuvierian organs absent.

Calcareous parts: Calcareous ring simple, not made up of a mosaic of smaller pieces (Figure 10). Ossicles tables, sometimes C-shaped rods, perforated plates.

Parastichopus californicus (Stimpson, 1857)

Common names: California Sea Cucumber, Giant Red Cucumber
Originally as: *Holothuria californica* Stimpson
Revised to: *Parastichopus californicus* by Deichmann (1937)
californicus = first described from California

Description

Parastichopus californicus can measure up to 50 cm long; it is the largest of the sea cucumbers in British Columbia waters. It is cylindrical with slightly tapered ends. The dorsal side has about 40 large, and many smaller, flesh-coloured papillae. The skin colour varies from mottled brown to a more solid brown or red in juveniles; all-white specimens occur rarely. The ventral side of the body has numerous rows of robust tube feet, and is usually lighter than the dorsal side. A circle of 20 peltate feeding tentacles surrounds the subterminal mouth (at the end but pointing down). See Photo 1.

Skin ossicles: tables with a disc diameter of 72 to 92 μm and a spire with 11 to 19 spines; large oval plates with two rows of holes running lengthwise.

Fig. 9. *Parastichopus californicus* in Saanich Inlet, B.C.

Similar Species

Parastichopus californicus might be confused with *P. leukothele*. In life, *P. leukothele* is quite distinct with bright orange skin and rusty-brown patches. The papillae are small and white. It also lives at greater depths, and only occasionally would be seen by scuba divers. In a preserved specimen the differences are less obvious, and one needs to make an ossicle preparation and count the number of spines on the table.

Synallactes challengeri is also similar in appearance, but it is smaller and differs in colour and in the length of the dorsal papillae. It is usually grey with a purple or pinkish tinge, and the papillae are longer and more slender.

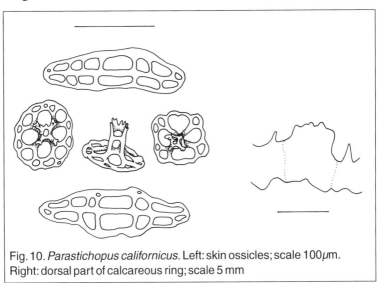

Fig. 10. *Parastichopus californicus.* Left: skin ossicles; scale 100µm.
Right: dorsal part of calcareous ring; scale 5 mm

Range

Gulf of Alaska to Cedros Island, west of Lower California; intertidal to 249 m.

Habitat

This species is common on mud, gravel, shell, rock rubble or solid bedrock, and from exposed coast to sheltered inlets. Reaches greatest densities in quieter waters where organic sediments settle on hard surfaces. In southern B.C., juveniles settle among dense mats of filamentous red algae and tubes of polychaete worms.

Biology

Parastichopus californicus feeds as it moves randomly along the bottom; in one study, *P. californicus* moved up to 3.9 metres in a day. The mop-shaped tentacles of *P. californicus* are pressed onto the substratum and small particles adhere to the sticky surface. As the tentacle retracts toward the mouth, the edges curl in to grasp larger particles. The tentacle enters the mouth and releases the particles. The animal gets its nutrition from the organic material in the sediment, mostly bacteria and fungi. As with many sea cucumbers, *Parastichopus californicus* ceases feeding and becomes dormant from September to early March.

The separate sexes reach maturity after four years. They are reported to migrate to shallow water to spawn from late April to August. In Friday Harbor, Washington, most spawning is from late May to mid July. During spawning, the anterior third to half of the body rises off the bottom in a cobralike posture. Strings of white sperm or light orange eggs (mean diameter = 204 μm) pour from the gonopore just behind the dorsal tentacles. Fertilization takes place in open water and the free-swimming larva feeds on plankton for 35 to 52 days before settling to the bottom.

Few predators are known — although *Parastichopus californicus* reacts strongly to the Sunflower Star (*Pycnopodia helianthoides*). When touched by this sea star, *P. californicus* rears back and flexes violently, effectively escaping from the grasp of the sea star.

From late summer to March most specimens have no obvious internal organs. Early workers thought that the internal organs were eviscerated through the cloaca. Recent evidence, however, indicates that the animal resorbs the organs during the dormant phase, and regenerates them during the winter.

Several parasites and commensals have been reported for *Parastichopus californicus*. The flatworm, *Anoplodium hymanae*, inhabits the body cavity; another flatworm, *Wahlia pulchella*, is found in the upper intestine; the parasitic gastropod, *Enteroxenos parastichopoli*, attaches to the intestine as elongated coils; the parasitic snail, *Vitriolina columbiana*, attaches to the external surface and penetrates the skin to suck the internal fluids and the commensal scale worm, *Arctonoe pulchra*, crawls on the skin among the papillae or tube feet.

Parastichopus californicus is harvested commercially in British Columbia, Washington and Alaska. In 1987-88 the volume of landings reached a peak of 1922 tonnes of whole animals in B.C. A coast-wide quota of 1500 tonnes was set from 1986-88, an area quota system from 1989 until 1995 and an Individual Quota system (IQ) was introduced in 1995. The total quota has been steadily reduced from the initial 1500 tonnes to 514 tonnes in 1995. In 1992 the landed value of the harvest was $1,363,000.00 and the wholesale value was $5,656,000.00. In 1994 the actual landings were 536 tonnes with a landed value of $976,000.00, based on an average price of $4.97 per kilogram of meat. The introduction of the IQ in 1995 may further reduce the total landings and help to safeguard the resource from overfishing.

References

Barr and Barr (1983), Black (1954), Burke et al. (1986), Cameron (1985, 1986), Cameron and Fankboner (1984, 1986), Da Silva et al. (1986), Deichmann (1937), Fankboner and Cameron (1985), Hufty (1973), Hufty and Schroeder (1974), Jones (1963), Leighton (1988), Levin et al. (1986), Lutzen (1979), Margolin (1976), Morris (1995), Shinn (1983; 1986), Sloan (1984, 1986a, 1986b), Smiley (1984, 1986a, 1986b, 1988a, 1988b, 1994), Smiley and Cloney (1985), Smiley et al. (1991), Smith (1966), Strathmann (1978), Strathmann and Sato (1969).

Parastichopus leukothele Lambert, 1986

leukothele = white papillae

Description

Parastichopus leukothele can measure up to 38 cm long (preserved). The elongate cylindrical body has a blunt anterior end and a tapering tail. Small (less than 1 cm), white papillae are scattered over the dorsal surface: approximately 2 per cm^2. The skin is rich orange with rusty brown patches around the bases of some papillae. The tube feet are confined to the ventral surface in four bands, each consisting of four rows. Twenty peltate feeding tentacles in two circles surround the mouth. The tube feet are usually white and often tipped with orange. See Photo 2.

Skin ossicles: tables and oval perforated plates; tables have mean diameter of 113 μm and 2 to 10 spines at the top of the spire.

Fig. 11. *Parastichopus leukothele* in Tasu Sound, B.C.

Similar Species

Parastichopus leukothele is similar in shape to *P. californicus*, but different in body colour and in the size of the papillae. *P. leukothele* has larger table ossicles, but with fewer spines on the spire.

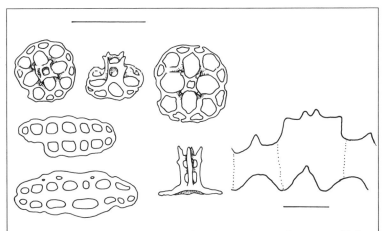

Fig. 12. *Parastichopus leukothele.* Left: skin ossicles; scale 100 μm. Right: dorsal part of calcareous ring; scale 5 mm.

Range

Recorded from Tasu Sound, Queen Charlotte Islands, to Point Conception, California; but, judging by their numbers at these locations, they very likely range farther north and south; from 24 to 285 metres depth.

Habitat

Most specimens are trawled from soft sediments on the continental shelf. Some have been observed from a submersible on sediment-covered rock ledges. I collected the type specimen at scuba-diving depth (26 m) on a rocky substratum in the Queen Charlotte Islands; but the species is rarely seen this shallow. In the past, this species has probably been lumped together with *P. californicus*, thus it is difficult to estimate its abundance from the literature. However, in Queen Charlotte Sound a 15-minute haul with a 10-foot otter trawl yielded 14 specimens: so in certain localities it is no doubt common.

Biology

Specimens in aquaria exhibit typical feeding behaviour, using their feeding tentacles like mops to pick up bottom sediments. Little is known about their reproduction except that eggs in the ovary are largest in March and April, and smallest from June to September. This suggests that spawning takes place in May. Unlike *P. californicus*, *P. leukothele* did not exhibit an escape response when touched by the Sunflower Star (*Pycnopodia helianthoides*).

The scale-worm, *Arctonoe pulchra*, occurs on the external surface; and I have found it in the body cavity of two specimens that had expelled their internal organs. Presumably, the scale-worm was able to enter the cavity through the damaged cloacal opening after auto-evisceration.

References

Lambert (1986, 1987, 1990a).

Family Synallactidae

Pseudostichopus mollis Théel, 1886
Synallactes challengeri (Théel, 1886)

External Features: Body with tube feet in rows ventrally, papillae dorsally. Body wall soft and pliable. Twenty equal, peltate tentacles. Internal Features: Tentacle ampullae absent. Retractor muscles absent. Rete mirable absent. Posterior mesentery attached to right ventral body wall. Gonad single tuft, or double tuft. Cuvierian organs absent.

Calcareous parts: Calcareous ring simple; not a mosaic of smaller pieces. Typical skin ossicles: tables or C-shaped bodies.

Pseudostichopus mollis Théel, 1886

mollis   =   soft

Description

Pseudostichopus mollis grows up to 15 cm in length. It is shaped like a flattened sausage. The skin is beige in colour and smooth; it is often covered by particles of sand or fine debris. Although this species has tube feet, they are so fine and sparse that they are not apparent to the naked eye. At the posterior end, the anus is in a vertical groove. The mouth is on the underside of the anterior end. The 20 peltate tentacles are fairly small and usually retracted inside the mouth. See Photo 3.

Few, if any, ossicles can be found in this species, thus only the calcareous ring is illustrated.

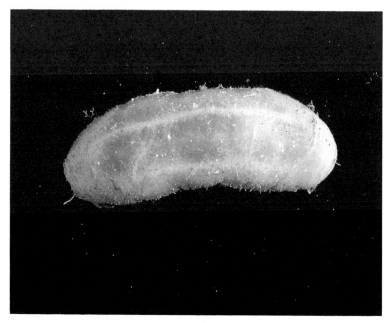

Fig. 13. Live *Pseudostichopus mollis* from Queen Charlotte Sound, B.C.

Similar Species

It is unlikely that *Pseudostichopus mollis* will be confused with any other species on the continental shelf.

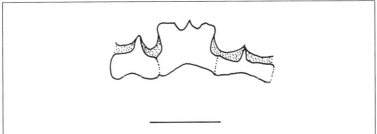

Fig. 14. *Pseudostichopus mollis.* Dorsal part of calcareous ring; scale bar 5 mm.

Range

In the northeastern Pacific it is known from the Gulf of Alaska to Oregon, from 179 to 2200 m; so it is only on the deeper part of the continental shelf along the coast. It was first collected and described from the South Atlantic and off Chile. Ludwig (1894) collected this species off the Galapagos Islands. It is a widespread species from deep water.

Habitat

Pseudostichopus mollis is dredged from mud or sand.

Biology

We know little about the biology of this species. Dr Frank Rowe (personal communication) reported that he saw hundreds of juveniles among the tentacles of one of the type specimens in the British Museum.

References

Alton (1972), Carney and Carey (1976), Clark, H.L. (1913, 1920, 1923), Ekman (1925), Ludwig (1894), Madsen (1953), Théel (1886).

Synallactes challengeri (Théel, 1886)

Originally: *Stichopus challengeri* Théel
Revised to: *Synallactes challengeri* by Ostergren (1896)
challengeri = from the *Challenger* expedition

Description

Synallactes challengeri can grow up to 20 cm in length. Its elongate, cylindrical body tapers toward the posterior end. The dorsal side of the body is covered with long, slender, tapering papillae. They do not appear to be in rows. The animal is grey with a tinge of pink or purple. On the ventral side, there are three series of tube feet. The middle series has two to four rows; the lateral series has only two. Twenty yellowish peltate tentacles surround the subterminal mouth. See Photo 4.

Skin ossicles: tables are a cross with a tall spire rising from the centre; rods (not shown) are long and slightly curved with racquet-shaped ends.

Ostergren (1896) placed this species in *Synallactes*. Clark (1922) also suggested that this species was not a *Stichopus,* but that it possessed the characters of the family Synallactidae.

Fig. 15. Live *Synallactes challengeri* from Queen Charlotte Sound, B.C.

Similar Species

Synallactes challengeri might be mistaken for a juvenile of a *Parastichopus*. The body colour, and size and shape of the papillae are fairly distinctive. Internally, *S. challengeri* has no tentacle ampullae, and the ossicles are quite different.

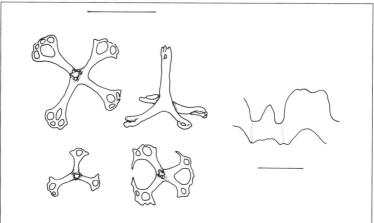

Fig. 16. *Synallactes challengeri.* Left: tables; scale 100 μm. Right: dorsal part of calcareous ring; scale 5mm.

Range

(Originally described as *Stichopus challengeri* from the southern Atlantic.) In the north Pacific known from Kodiak Island, Alaska, to northern Vancouver Island; 20 to 366 metres depth.

Habitat

S. challengeri lives on a variety of substrata from rock to mud, usually with gravel. In Portland Inlet, northern B.C., I have seen it in shallow water on a rocky bottom.

Biology

This species feeds on bottom sediments with its peltate tentacles, like *Parastichopus*. We know little about the other aspects of its biology.

References

Clark (1922), Edwards (1907), Ostergren (1896), Théel (1886).

Family Psolidae

Psolidium bidiscum Lambert, 1996
Psolus chitonoides Clark, 1901b
Psolus squamatus (Koren, 1844)

External Features: Body domed dorsally with stiff, shinglelike scales, flat, flexible sole ventrally. Tube feet may be scattered dorsally, in rows ventrally, or ventral series only. Tentacles 8-10, dendritic, equal in size, or 8 large and 2 small.

Internal Features: Tentacle ampullae absent. Retractor muscles present. Respiratory trees Y-shaped. Rete mirable absent. Gonad double tuft. One madreporic body, attached to dorsal mesentery. Cuvierian organs absent.

Calcareous parts: Calcareous ring with anterior processes only. Not a mosaic. Typical skin ossicles: perforated plates, and baskets or cups.

Psolidium bidiscum Lambert, 1996

bidiscum = referring to two types of plates in the sole

Description

Psolidium bidiscum was misidentified for many years as *Psolidium bullatum* Ohshima, but a detailed examination showed it to be an undescribed species. *P. bullatum* is known only from the Aleutian Islands and Gulf of Alaska. *P. bidiscum* is a small (1 to 3 cm), dome-shaped sea cucumber with mauve to pink, shinglelike plates on the top side; about 10 rows of scales between the mouth and the anus. Each plate has up to 13 knoblike bumps and up to 6 tiny tube feet protruding through pores in the plate. The flat sole has three rows of tube feet around the edge, made up of two rows of robust tube feet and one outer row of minute tube feet. A staggered row of tube feet runs down the centre of the sole. The 10 feeding tentacles, 8 equal and 2 smaller, are translucent white, with reddish-brown blotches. Many specimens have prominent reddish-brown bands near the bases of the two smaller tentacles. See Photo 5.

Skin ossicles: two sizes of perforated plates in the ventral sole and cups in the dorsal membrane that covers the plates.

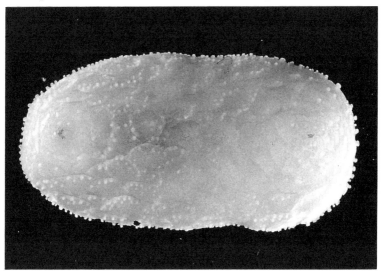

Fig. 17. Preserved *Psolidium bidiscum* dredged from Prevost Passage, B.C.

Similar Species

There are two species similar to *Psolidium bidiscum* in our waters: *Psolus chitonoides* and *Psolus squamatus*. The most common, *P. chitonoides*, is bright red to orange, with red tentacles; it does not have minute tube feet piercing the dorsal scales. *P. squamatus* has white dorsal scales and white tentacles. *Psolidium bidiscum* and *Psolus squamatus* both occur in deeper water. *P. bidiscum* is also accessible by scuba, but the sediment-covered body is difficult to spot.

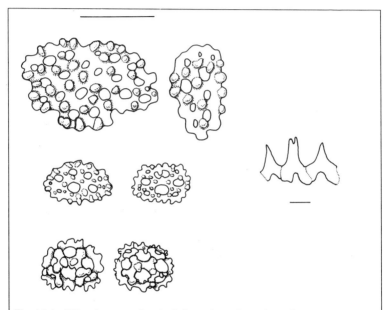

Fig. 18. Left Top: large perforated plates from the sole. Left Middle: small plates from the sole. Left Bottom: cups from the dorsal membrane. Right: dorsal part of calcareous ring. Left scale 100 μm; right scale 1 mm.

Range

Southeastern Alaska to central California; intertidal to 220 metres, but most commonly in 20 to 60 m.

Habitat

Psolidium bidiscum is known mostly from dredged specimens attached to rocks or shells. However, I have seen it at 10 m in

Saanich Inlet, British Columbia, on sediment-covered rock surfaces. The pale tentacles are exposed, but sediment usually covers the body. When the tentacles are retracted it can easily be overlooked. It may be more common in shallow water than the number of specimens in collections suggests, because of the difficulty in sampling solid rock with a dredge. In San Juan Channel, Washington, it is found in densities of 10 per 0.1 square metres. It has only been recorded inter-tidally in one locality in Hood Canal, Puget Sound.

Biology

Psolidium bidiscum feeds like *Psolus chitonoides*. Sticky tentacles catch particulate matter and are placed into the mouth individually and cleaned off.

Spawning occurs from late March to early May. The anterior end of the animal lifts slightly off the substratum during spawning. The female may spawn up to 3,000 golden yellow to light brown-orange eggs (mean diameter 330 μm). When ripe, the tan ovary or white testis is visible through the thin ventral sole. The larva is a pelagic lecithotrophic vitellaria with three ciliary rings.

References

Austin (1985), Bakus (1974), Hadfield (1961), Hetzel (1960), Lambert (1996), McEuen (1987, 1988), McEuen and Chia (1991), Stricker (1986).

Psolus chitonoides Clark, 1901

Common names: Armoured Sea Cucumber, Creeping Armoured Cucumber, Slipper Sea Cucumber, Creeping Pedal Cucumber

chitonoides = like a chiton

Description

Psolus chitonoides is an unusual species that resembles a chiton rather than a typical sea cucumber. It is dome-shaped, and covered with pale yellow to bright orange, shinglelike calcareous plates; about four rows between the mouth and anus. The sole is soft, flat and pale orange to white. The body is oval, and can grow up to 7 cm long and 5.8 cm wide. A plume of 10 equal-sized, bright red, white-tipped tentacles, protrudes through an opening in the dorsal plates at one end. The anus exits through an opening at the opposite end on the dorsal side. Tube feet are confined to the sole. A series of two to four rows of robust tube feet run down each side of the sole; and another row of tiny tube feet in small depressions run down the outer edge of the sole. A staggered series of tube feet runs down the centre of the sole. See Photo 6.

Skin ossicles: one type of circular perforated plate — some with knobs coalesced into a raised network — occur only in the ventral sole.

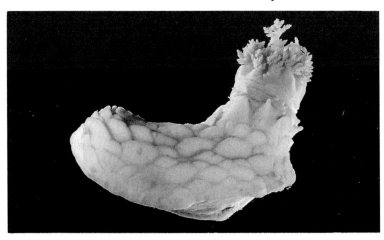

Fig. 19. Preserved *Psolus chitonoides* from North Pender Island, B.C.

Similar Species

There are two species similar to *Psolus chitonoides*: *Psolidium bidiscum* and *Psolus squamatus*. *P. bidiscum* is smaller (1 to 3 cm), and is normally dredged from deeper water. It has tube feet piercing the dorsal scales; and the tentacles are pale pink or white, as opposed to the bright red, white-tipped tentacles of *P. chitonoides*. *Psolus squamatus* has white scales and is dredged from deeper waters.

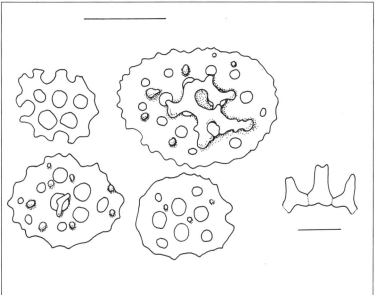

Fig. 20. *Psolus chitonoides.* Left: ossicles from ventral sole; scale bar 100 μm. Right: dorsal part of calcareous ring; scale 2mm.

Range

Aleutian Islands to Baja California; intertidal to 247 m; common in shallow subtidal areas.

Habitat

Psolus chitonoides occurs in a range of habitats from exposed coast to sheltered inlets; although it seems to prefer clean, vertical rock that is free of sediment. Its soft, flat sole enables it to attach firmly to rock. Other organisms attach to the dorsal side, often only the tentacles indicating its location.

Biology

Psolus chitonoides is a suspension feeder. The tentacles trap larger particles (greater than 2 mm) by bending inwards to form a cagelike enclosure. The mouth lips extend toward the particle as the nearest tentacle pushes it into the mouth.

Spawning occurs from mid March to late May, commonly in the early morning. A spawning male will swab its genital papilla with its tentacles, then lift the tentacles to disperse the sperm. Females release long ropes of brick red eggs (mean diameter 627 μm). Fertilized eggs develop into pelagic lecithotrophic vitellaria larvae (Fig. 8C). Late larvae and early juveniles are negatively phototactic and settle gregariously.

Toxic chemicals (saponins) discourage fish from nipping the tentacles. Even the Kelp Greenling (*Hexagrammos decagrammus*) — which commonly feeds on sea cucumbers — avoids *P. chitonoides*. The Sun Star (*Solaster stimpsoni*), the Leather Star (*Dermasterias imbricata*), the Sunflower Star (*Pycnopodia helianthoides*) and the Red Rock Crab (*Cancer productus*) prey on *P. chitonoides*.

References

Bakus (1974), Bingham and Braithwaite (1986), Chia and Burke (1978), Chia and Spaulding (1972), Chia et al. (1975), Clark, H. L. (1901a, 1901b, 1924), Dybas and Fankboner (1986), Emlet (1982), Engstrom (1974), Fankboner (1978), Gotshall and Laurent (1979), Hetzel (1960), Jones (1962), Mauzey et al. (1968), McDaniel (1973), McEdward and Chia (1991), McEuen (1987, 1988), McEuen and Chia (1991), Ohshima (1915), Young and Chia (1982).

Psolus squamatus (Koren, 1844)

Originally as: *Cuvieria squamatus* Koren
Revised to: *Psolus squamatus* by Lütken (1857)
squamatus = scaly

Description

Psolus squamatus has the typical psolid shape: a low hemisphere with a flat base. It can grow up to 13 cm long, but normally ranges between 5 and 10 cm. The dorsal scales are white and smooth, but often discoloured brown. Between the tentacles and the anus there are about 12 plates. The thin, dorsal plates overlap smoothly but some have granules along their edges. The tube feet form a conspicuous double row around the perimeter of the sole with a few scattered down the mid-line. The 10 white tentacles are approximately equal in size. See Photo 7.

Skin ossicles: only sparse netlike perforated plates in the sole.

Similar Species

Psolus squamatus can be distinguished from the other two psolids by its colour: *P. squamatus* is white; *P. chitonoides* is red and *P. bidiscum* is mauve-pink.

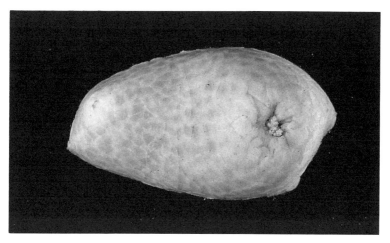

Fig. 21. Preserved *Psolus squamatus* dredged off southwest coast of Vancouver Island.

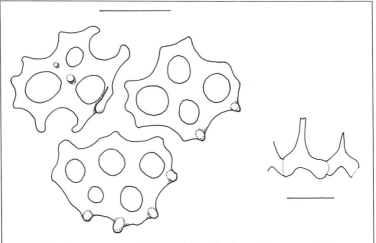

Fig. 22. *Psolus squamatus*. Left: ossicles of sole; scale 100 μm. Right: dorsal part of calcareous ring; scale 5mm.

Range

A widespread species known from the southern Bering Sea to Cape Horn, South America and from the British Isles and Norwegian coast. In B.C. it has been collected between 37 and 1061 m.

Habitat

As with the other psolids, this species usually attaches to solid rock, stones, or mollusc shells. Thus, its true abundance is difficult to determine using conventional dredging gear. However, using the *Pisces* submersible, I have observed numerous individuals on vertical rock walls in several coastal inlets.

Biology

P. squamatus is a typical suspension feeder. Little is known about reproduction or development in this species. It is reported to be sexually mature at a length of 30 mm.

References

Alton (1972), Clark, H. L. (1901a, 1923), Deichmann (1941, 1947), Ekman (1923), Imaoka (1980), Lambert (1984), Ludwig (1900), Mortensen (1977), Östergren (1902), Perrier (1905).

Family Cucumariidae

Subfamily Cucumariinae
Cucumaria frondosa japonica (Gunnerus, 1767)
Cucumaria miniata (Brandt, 1835)
Cucumaria pallida Kirkendale and Lambert, 1995
Cucumaria piperata (Stimpson, 1864)
Cucumaria pseudocurata Deichmann, 1938b
Cucumaria vegae Théel, 1886
Pseudocnus curatus (Cowles, 1907)
Pseudocnus lubricus (H.L. Clark, 1901b)

Subfamily Thyonidiinae
Ekmania diomedeae (Ohshima, 1915)
Thyonidium kurilensis (Levin, 1984)

External Features: Body cucumber-like. Body wall soft and pliable. Tube feet in five series, or scattered all over body. Tentacles 10-20. Tentacles dendritic, equal in size, or 8 large and 2 small.

Internal Features: Tentacle ampullae absent. Retractor muscles present. Respiratory trees Y-shaped. Rete mirable absent. Gonad double tuft. Madreporic bodies 1-50: one is attached to dorsal mesentery; accessory madreporites around water ring. Cuvierian organs absent.

Calcareous parts: Calcareous ring with anterior processes only. Not a mosaic. Typical skin ossicles perforated plates or buttons. Large complex bodies pine-cone shaped.

Cucumaria frondosa japonica (Gunnerus, 1767)

frondosa	=	**full of leaves**
japonica	=	**refers to Japan**

Description

Cucumaria frondosa japonica is the largest of the *Cucumaria* species on our coast. It has the typical *Cucumaria* shape (Fig. 1A), but has a grey to black body. The tentacles and mouth region can be quite colourful, with areas of white or red. When dredged or disturbed it resembles a grey football. It is gigantic compared to other species of *Cucumaria* on this coast. It can attain a length of 30 cm and longer, depending on its state of relaxation. The contractile tube feet are in five rows. The 10 equal-sized tentacles are black and bushy when extended. See Photo 8.

Skin ossicles: quite variable and usually cannot be found in large specimens; in small specimens, elongate perforated plates with large holes, sharp bumps on the surface and scalloped margins. In the introvert, the ossicles are larger and more complex.

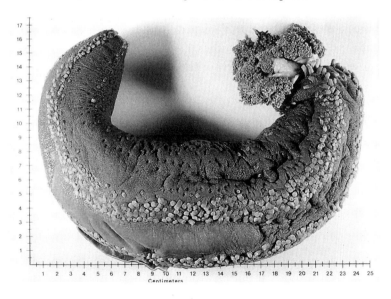

Fig. 23. Preserved *Cucumaria frondosa japonica* from Pearse Island, B.C.

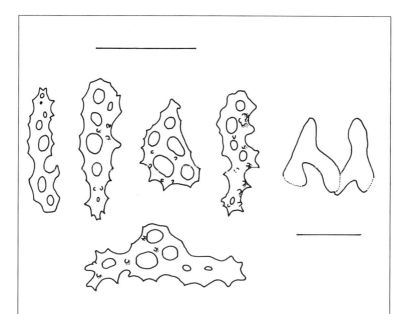

Fig. 24. *Cucumaria frondosa japonica.* Left: skin ossicles; scale 100 μm. Right: dorsal part of calcareous ring; scale 5mm. (From a 6-cm specimen.)

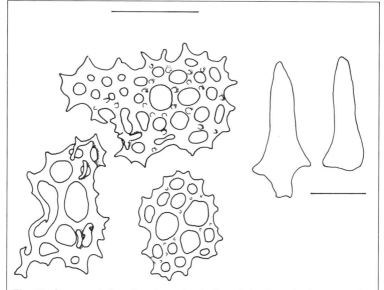

Fig. 25. *Cucumaria frondosa japonica.* Left: ossicles from the introvert of a large specimen (18 cm). Right: dorsal part of calcareous ring. Ossicles do not appear in the general body wall. Left scale 100 μm, right scale 5 mm.

Similar Species

Cucumaria frondosa japonica resembles *Cucumaria miniata* in general body form, but the former is usually larger and darker. A juvenile *C.f. japonica* might be confused with *C. miniata*; however, the latter is usually orange. *Cucumaria frondosa* is a common sea cucumber on the east coast of North America. *Cucumaria japonica* is the closely related species in Japanese waters of the northwestern Pacific. Some consider *japonica* to be a subspecies of *frondosa,* but there is presently no agreement. The specimens on this coast seem to be intermediate in many characters, hence I have used the subspecies designation. However, Dr V. Levin from Russia has looked at our specimens and has concluded that they are not *C. f. japonica* and suggests there may be more than one species. Until the status of this species is resolved, I shall use the name *C. f. japonica* to indicate its close relationship to *C. frondosa* and *C. japonica*.

Range

Alaska to Fitzhugh Sound, near Bella Bella, B.C.; 25 to 130 metres. Reported in the Port Hardy region of northern Vancouver Island, but this has yet to be confirmed.

Habitat

Cucumaria frondosa japonica is uncommon in the south. It is usually dredged from sand or gravel, but can also be collected by scuba from rock substrata. It has been reported from shallow water in Burke Channel in large numbers on a rocky substratum.

Biology

There are no studies on the biology of this subspecies on the west coast of North America. Because it is so closely allied to *C. frondosa* on the east coast, the biology of that species could be used as a guide (see Jordon, 1972).

References

Chia and Burke (1978), Chia et al. (1975), Clark, A. H. (1920), Duncan and Sladen (1881), Edwards (1907, 1910a, 1910b),

Filimonova and Tokin (1980), Jordan (1972), Klugh (1923), Krishnan and Dale (1975), Ludwig (1900), Mortensen (1932), Mottet (1976), Panning (1955), Pawson (1977), Runnstrom and Runnstrom (1921), Selenka (1867), Shick (1983), Sutterlin and Waddy (1976).

Cucumaria miniata (Brandt, 1835)

Common names: Orange Sea Cucumber, Red Sea Gherkin
Originally as: *Cladodactyla miniata* Brandt
Revised to: *Cucumaria miniata* by Selenka (1867)
miniata = bright red

Description

Cucumaria miniata is a sausage-shaped sea cucumber that grows up to 15 to 20 cm long. It has five bands of tube feet separated by broad expanses of smooth skin, and occasional scattered tube feet. Its colour varies from light orange to a dark orange-brown. The introvert is usually a lighter shade than the rest of the body. The 10 equal-sized tentacles are quite full and bushy when they are extended. Tube feet on the introvert are more prominent than on the body. See Photo 9.

Skin ossicles: flat, oval perforated plates, slightly tapered and seldom round; with occasional, pointed bumps on the surface of the plate.

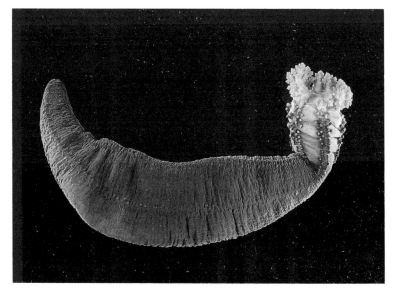

Fig. 26. Preserved *Cucumaria miniata* from Sooke, B.C.

Similar Species

Cucumaria pallida is a pale orange or white sea cucumber that lives in the same habitat as *Cucumaria miniata,* and was previously thought to be a pale form of this species. The 10 white tentacles of *C. pallida* are thin and wispy compared to the robust, bushy orange or brown tentacles of *C. miniata.*

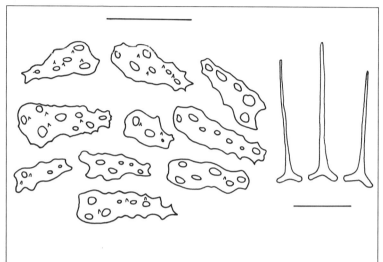

Fig. 27 *Cucumaria miniata.* Left: skin ossicles; scale 100 μm. Right: dorsal part of calcareous ring; scale 5 mm.

Range

Amaknak Island, Aleutian Islands, to Cambria, California (this study); south to San Benito Island (Bergen 1996); intertidal to 225 m, but most abundant from 0 to 25 m. Three records from deeper than 100 m seem exceptionally deep and may prove to be erroneous.

Habitat

Cucumaria miniata is a common intertidal sea cucumber on the coast. It is found in sheltered waters among rock rubble in low intertidal or shallow subtidal areas. The body is hidden under rocks, but the bright orange or brown tentacles extend up into the water. This species is especially abundant where tidal current and rock rubble occur together.

Biology

Cucumaria miniata exhibits typical suspension-feeding behaviour. (See Page 6.) Feeding tentacles withdraw rapidly when touched. *C. miniata* feeds very little from November to March, when the plankton is reduced. Most specimens lie hidden beneath the rocks. The posterior end protrudes from beneath the rock to fill the respiratory trees with freshly oxygenated water.

Spawning is from early March to mid May, during intervals of slack tide. The anterior part of the body extends and green eggs (mean diameter 520 mm) are released in buoyant pellets which later fall apart. The yolky egg develops into a sluggish, non-feeding larva. Juveniles settle onto the undersides of rocks near the adults.

Predators of *Cucumaria miniata* include the Sand Star (*Luidia foliolata*), the Northern Sun Star (*Solaster endeca*) and the Sun Star (*Solaster stimpsoni*). The latter causes *C. miniata* to react with violent contractions in an effort to escape. The Kelp Greenling nips off the tentacles of this sea cucumber.

Like vertebrates, *Cucumaria miniata* uses haemoglobin to absorb and transport oxygen to the cells. The blood cells of *C. miniata* also show many structural similarities to the blood cells of some lower vertebrates such as trout and hagfish.

Some specimens of *Cucumaria miniata* contain a parasitic gastropod, *Thyonicola dogieli.* This parasite appears as a long egg-filled tube that is coiled like a spring.

References

Bakus (1974), Barr and Barr (1983), Bergen (1996), Bingham and Braithwaite (1986), Britten (1906), Brumbaugh (1980), Byrne (1986), Chia and Spaulding (1972), Chia et al. (1975), Clark, H. L. (1901a, 1924), Edwards (1910a), Engstrom (1974), Fontaine and Lambert (1973, 1976, 1977), Gotshall and Laurent (1979), Hetzel (1960), Lutzen (1979), Manwell (1959), McEdward and Chia (1991), McEuen (1987, 1988), Morris et al. (1980), Mottet (1976), Pearse et al. (1988), Selenka (1867), Shimek (1987), Stricker (1986), Terwilliger (1975), Terwilliger and Read (1970), Tikasingh (1960), Wolcott (1981).

Cucumaria pallida Kirkendale and Lambert, 1995

pallida = pale

Description

Cucumaria pallida has a typical cucumarid shape. It can grow up to 26 cm long. This species has five rows of tube feet — the three ventral rows being better developed than the other two. In preserved specimens the tube feet may be retracted and appear as dimples. In a live specimen, the 10 equal-sized tentacles are long and wispy and usually white. The body is pale orange-white or tan. See Photos 10 & 11.

Skin ossicles: circular or oval perforated plates, some with one end tapering; knobs on surface of many plates.

Fig. 28. Live *Cucumaria pallida* from Saanich Inlet, B.C.

Similar Species

Until recently, *Cucumaria pallida* was assumed to be a pale form of *Cucumaria miniata*. They live in similar habitats and breed close to the same time. The most obvious difference, apart from the colour, is the form of the tentacles when extended. *Cucumaria pallida* has thin, wispy, white tentacles, whereas *Cucumaria miniata*'s tentacles are thicker at the base, bushier and usually orange or brown. Internally, *Cucumaria pallida* has an average of 9 madreporic bodies, while *Cucumaria miniata* has 45.

The white tentacles of *Eupentacta* can be confused with those of *C. pallida,* but the former has eight larger tentacles and two tiny ones. *C. pallida* has 10 equal tentacles.

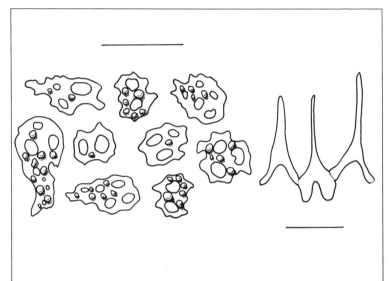

Fig. 29. *Cucumaria pallida.* Left: skin ossicles; scale 100 μm. Right: dorsal part of calcareous ring; scale 5 mm.

Range
Auke Bay, Alaska to Santa Rosa Island, California. No specimens have been found yet from the outer coasts of Washington, Oregon and northern California, but I would expect them to occur there. Known from the intertidal zone to a depth of 91 m. Half the records are from depths between 11 and 20 m.

Habitat
Cucumaria pallida is common beneath rock rubble in sheltered waters or in a current. It often occurs together with *Cucumaria*

miniata, but unlike the latter is more common in quieter waters. It is common in Saanich Inlet, British Columbia; and also occurs among rubble at Ogden Point Breakwater, Victoria.

Biology

Cucumaria pallida feeds in the same way as *C. miniata.*

Spawning occurs from mid March to early May. Adult females produce long strands of eggs, 1 to 2 eggs wide, which break up in 12 to 15 minutes. Up to 8,800 tan-coloured eggs (mean diameter 504 μm) are produced. Each egg develops into a pelagic three-ringed doliolaria larva, which is repelled by light.

Because this species was assumed to be *Cucumaria miniata,* there is little published information on the biology of this species. McEuen (1987) recognized it as a different species based on its reproductive biology, but referred to it as *Cucumaria fallax* in his publications.

References

Kirkendale and Lambert (1995), McEuen (1986, 1987, 1988).

Cucumaria piperata (Stimpson, 1864)

Originally as: *Pentacta piperata* Stimpson
Revised to: *Cucumaria piperata* by H.L. Clark (1901a)
piperata   =   peppered

Description

Cucumaria piperata is similar in body form to *Cucumaria miniata,* but smaller (up to 12 cm long). It is creamy white, or slightly yellowish, with a variable number of black or dark brown spots. The spots may cover the body, but are concentrated at the anterior end and on the 10 equal-sized tentacles. The body has five double rows of retractile tube feet, and the skin is smooth. See Photo 12.

Skin ossicles: tapered perforated plates with spiny margins.

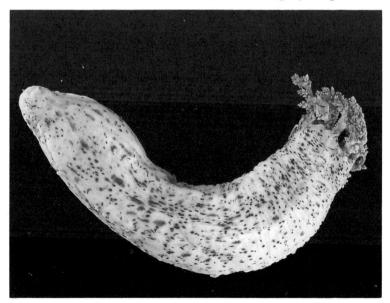

Fig. 30. Preserved *Cucumaria piperata* from Quatsino Sound, B.C.

Similar Species

Cucumaria piperata may be confused with *Pseudocnus lubricus*. One form of *P. lubricus* is yellowish white, with fine peppery spots on the dorsal side, and is shorter than 5 cm. Formerly, the speckled form of *P.*

lubricus was identified as *Cucumaria fisheri astigmata* or *Pseudocnus astigmatus*; but these names are synonymous with *Pseudocnus lubricus* — formerly *Cucumaria lubrica* (Arndt et al. 1996).

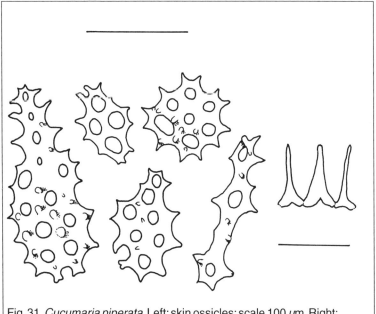

Fig. 31. *Cucumaria piperata*. Left: skin ossicles; scale 100 μm. Right: dorsal part of the calcareous ring; scale 5 mm.

Range

Known from the Queen Charlotte Islands south to Puget Sound, and also reported south to San Benito Island, Baja California (Bergen 1996). It is unclear whether some other reports from California are truly *C. piperata,* or the speckled form of *P. lubricus*. *C. piperata* is usually found in shallow subtidal, but is known as deep as 137 m.

Habitat

Cucumaria piperata is common in cobble fields, but less abundant than *Cucumaria miniata*. This species is found in a variety of habitats — firm mud, shell-gravel or hard surfaces that are exposed to current. It is usually buried or under rocks, but can also be found lying on the surface.

Biology

Cucumaria piperata is a suspension feeder. The tentacles are less robust than *Cucumaria miniata* and held closer to the substratum. Because of this feeding position and the mottled colour, *C. piperata* is more cryptic.

Spawning is similar to *C. miniata* and occurs from mid March until late April. The olive green eggs (mean diameter 532 μm) are packed into pellets which float when first released. The non-feeding larva is difficult to distinguish from that of *C. miniata*. Juveniles settle under rocks with adults of *C. piperata* and *C. miniata* in July.

No parasitic gastropods are reported in *C. piperata*. The body wall of *C. piperata* is toxic to the gunnels, *Pholis* and *Apodichthys*.

References

Arndt et al. (1996), Bakus (1974), Bergen (1996), Bingham and Braithwaite (1986), Brumbaugh (1980), Chia et al. (1975), Clark, A. M. and Rowe (1967), Clark, H. L. (1924), Deichmann (1937), Engstrom (1974), Hetzel (1960), Lambert (1984), McEuen (1987, 1988), Panning (1962), Shimek (1987), Stimpson (1864), Terwilliger and Read (1970).

Cucumaria pseudocurata Deichmann, 1938b

Common name: Tar Spot Sea Cucumber
pseudo   =   **false, refers to similarity to *Pseudocnus curatus***
curata   =   **to care for, refers to brooding of eggs**

Description

Cucumaria pseudocurata is a small species, averaging 1.5 to 3 cm in length. The dorsal side varies from brownish black to light brown to yellowish grey, the ventral side from brown to white. The five bands of tube feet are in single or zigzag rows — the three ventral rows being more robust. There are no tube feet scattered between the rows. As a rule, there are eight equal-sized tentacles and two smaller ventral ones; but occasionally populations have been reported with a large proportion of equal-sized tentacles. This may have been due to misidentification, however. The tips of the tentacles are usually the most darkly pigmented. A genital papilla occurs between two of the dorsal tentacles. See Photo 13.

Skin ossicles: oval perforated plates; typically with two central oval holes surrounded by smaller holes. In the southern part of *C. pseudocurata*'s range, its ossicles are large and oval, but towards the north, its ossicles are more narrow and smoother around the edge.

Fig. 32. Preserved *Cucumaria pseudocurata*.

Similar Species

In British Columbia *Cucumaria pseudocurata* might be confused with *Pseudocnus curatus* (formerly *Cucumaria curata*). *P. curatus* is usually black or dark brown, has 10 equal-sized tentacles, and tube feet scattered on the dorsal side. It is usually found at shallow subtidal or low intertidal — while *C. pseudocurata* is found at the mid intertidal level, near or among mussels.

To further confuse identifications, *Cucumaria pseudocurata* is hard to differentiate from *Cucumaria vegae* from Alaska. In fact, they appear identical to the naked eye. There is a gradual change in the ossicles of *C. pseudocurata* with latitude, but there is no clear distinction between it and *C. vegae*. Recent work with the DNA of these two species has confirmed that they are closely related, with only a 2% difference in the DNA between *C. pseudocurata* in the south and *C. vegae* in the north. In this type of analysis a difference of greater than 2% would indicate that the two forms were distinct

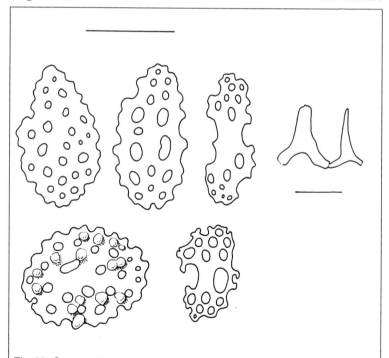

Fig. 33. *Cucumaria pseudocurata.* Left: skin ossicles; scale 100 μm. Right: dorsal part of the calcareous ring; scale 1 mm.

species, 2% or less is not sufficient to separate them. Because these two names are already in the literature, and it is still not clear whether to synonymize them, I have retained them as separate species in this book. For the time being, I would identify a specimen from south of the Queen Charlotte Islands as *C. pseudocurata* and those from the Queen Charlottes north to the Aleutians as *C. vegae*.

Range

DNA analysis suggests that specimens occurring from the central coast of British Columbia and northern Vancouver Island south to Monterey Bay, California should be considered as *C. pseudocurata*. Occur in the mid to low intertidal zone.

Habitat

Cucumaria pseudocurata is common in British Columbia. It is usually associated with the California mussel (*Mytilus californianus*) and, therefore, only found on the exposed coast. *C. pseudocurata* was seen near Bamfield on a semi-exposed rocky shore in great abundance among the alga *Rhodomela*, just below the mussel zone. Its distribution is patchy, being abundant in one part of the mussel bed and absent in another. In the southern part of the range, large aggregations inhabit the lower intertidal zone on the landward side of large rocks.

Biology

The mussel bed in which *C. pseudocurata* lives offers protection from the ocean swells. Its tentacles pick up various suspended particles to consume.

Unlike most sea cucumbers, *Cucumaria pseudocurata* spawns from mid December to mid January. Males raise the anterior end to release sperm bound in strings of mucus that sink to the bottom. The female spawns up to 340 bright orange eggs (mean diameter 1051 μm). She places the eggs between her ventral side and the substratum until they develop into crawling juveniles in two or three months. *Cucumaria pseudocurata* is reported to live for about five years, but is not reproductive until its third year.

The six-armed sea stars (*Leptasterias spp.*) are potential predators on *C. pseudocurata*. In California the Sunflower Star, *(Pycnopodia helianthoides)*, feeds on the lower edge of sea cucumber aggregations.

References

Arndt et al. (1996), Atwood (1975), Bakus (1974), Brumbaugh (1965, 1980), Cherbonnier (1951), Chia et al. (1975), Collison (1983), Deichmann (1938b), Lambert (1985), Lambert (in press), McEuen (1987, 1988), Nordhausen (1972), Rutherford (1973), Smith (1962), Turner and Rutherford (1976).

Cucumaria vegae Théel, 1886

vegae   =   **possibly from Latin, *vagus*, meaning wandering**

Description

Externally, *Cucumaria vegae* is identical to *Cucumaria pseudocurata,* but often a bit larger. The difference between them can only be determined by examining their skin ossicles. Even then, specimens from southeastern Alaska have ossicles that are intermediate between those from the Aleutian Islands and southern British Columbia. Analysis of mitochondrial DNA has shown that specimens occurring from the Queen Charlotte Islands to the Aleutians form a group that is about 2% different from the southern population we call *Cucumaria pseudocurata,* that lives from Vancouver Island south to California. For the time being, I will retain the existing names and propose that the name *Cucumaria vegae* be used for all specimens of this description found north of Queen Charlotte Sound. See Photo 14.

Skin ossicles: vary from simple rods with a few holes, to rods that are expanded at the ends.

Fig. 34. Preserved *Cucumaria vegae* from Cook Inlet, Alaska.

Similar Species

As explained above, *Cucumaria pseudocurata* and *C. vegae* are similar and may not be a separate species. *Pseudocnus curatus* differs in the arrangement of its tube feet and the equal size of its tentacles.

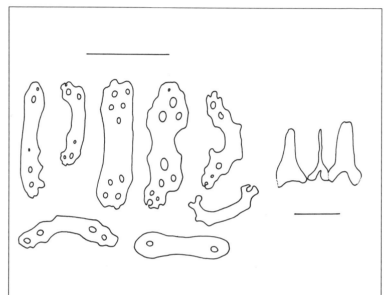

Fig. 35. *Cucumaria vegae*. Left: skin ossicles; scale 100 μm. Right: dorsal part of the calcareous ring; scale 1 mm.

Range

Commander Islands, at the western end of the Aleutian Islands south to the Queen Charlotte Islands, British Columbia. Also south to Hokkaido, Japan, in the western Pacific.

Habitat

Cucumaria vegae inhabits rocky intertidal areas among the California Mussel *(Mytilus californianus)* when that species is present: otherwise, it forms aggregations on rock surfaces in the lower intertidal. Near Juneau, Alaska, this species occurs in dense mats in the lower intertidal along with the White Sea Cucumber, *Eupentacta pseudoquinquesemita.*

Biology
Very little has been published about *Cucumaria vegae*. So far all we know is that it broods its young in late winter.

References
Barr and Barr (1983), H.L. Clark (1902, 1924), D'jakonov (1949), Edwards (1907), Lambert (in press), Mitsukuri (1912), Panning (1955), Ricketts et al. (1985), Stickle and Denoux (1976).

Pseudocnus curatus (Cowles, 1907)

Originally as: *Cucumaria curata* Cowles
Revised to: *Pseudocnus curatus* by Lambert (in press)
curata = to care for, referring to its brooding behaviour

Description

Pseudocnus curatus is a small, black sea cucumber about 1 to 3 cm in length. Its mouth and anus are slightly upturned; and its body tends to be a bit flattened. The tube feet are like scattered pits on the dark, dorsal surface, rather than being in rows. Ten equal-sized tentacles surround the mouth. The lighter-coloured ventral tube feet are usually more robust and in rows. The dorsal colour is a solid black or dark brown. See Photo 15.

Skin ossicles: four-holed buttons with smooth wavy edges, some with more than four holes. Three-armed rods in the tube feet.

Fig. 36. Preserved *Pseudocnus curatus* from Esquimalt, B.C.

Similar Species

The identity of *Pseudocnus curatus* has been confused for a number of years. This species was described from California and until

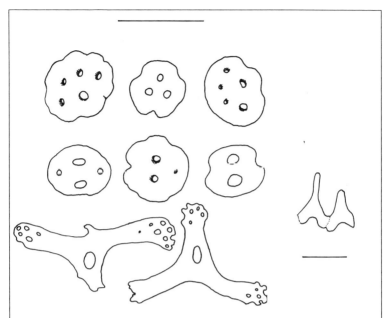

Fig. 37. *Pseudocnus curatus*. Left: skin ossicles; scale bar 100 *µ*m. Right: dorsal part of the calcareous ring; scale 2 mm.

recently was thought to be found only around the Monterey region. It has been confused with *Cucumaria pseudocurata,* which is also black and intertidal. The latter has rows of tube feet on the dorsal side rather than being scattered; and it usually has eight large tentacles and two small ventral ones. Both species can occur on the same beach within a few feet of each other, so are often confused. To be certain of identification, one must look at the ossicles in the skin. Compare the ossicles of *P. curatus* with those of *C. pseudocurata.* The former has compact four-holed buttons and three-armed rods, the latter has oval, perforated plates.

 P. curatus (=C. curata) has also been confused with *P. lubricus (=C. lubrica).* Most references identify the small black aggregating species found subtidally in high current areas of Juan de Fuca Strait as *C. lubrica.* This appears now to be incorrect (Arndt et al. 1996). Studies of the two species' ossicles and most recently, mitochondrial DNA, have shown that what used to be identified as *C. lubrica* is actually *P. curatus.* The true *P. lubricus* is also a brooding species that aggregates, but usually varies from white to brown with black specks on the dorsal side. Do not confuse it with *Cucumaria piperata,* which has black spots all over its body. *P. curatus* has smooth buttons while *P. lubricus* has more bumpy, lobed buttons. These two species are closely related and show about a 7% difference in their mitochondrial DNA.

Range

Because of the confusion with its identification, we cannot rely on the literature to determine the range. There are specimens of *P. curatus* in the RBCM collection from as far north as the B.C.-Alaska border, outer coast of Vancouver Island and in the Victoria region. It probably occurs from northern B.C. south to central California, but there are many gaps such as on the outer coast of Washington and Oregon that need to be checked. It occurs from the low intertidal down to about 20 m.

Habitat

Pseudocnus curatus occurs on rocky intertidal and shallow subtidal rock in strong current or open coast surf. In California I have found it nestled in crevices in small groups sometimes partly overgrown with encrusting coralline algae on exposed rock. It may occur subtidally there, but perhaps few people have looked for this species in those exposed locations. In Juan de Fuca Strait it tends to be shallow subtidal in areas where the tidal currents are strong. They nestle together in large aggregations on open rocky surfaces.

Biology

Cowles (1907) described many individuals forming black patches just below the low tide mark, clinging tightly to the rocks. As soon as the eggs are laid, the mother transfers them to the ventral surface: here, they develop directly into juveniles. The eggs are about 1 mm in diameter when laid. Associated with this species during the breeding season is a small nematode that feeds on the eggs, often destroying the whole brood. A paper by Engstrom (1974) on the biology of *C. lubrica* probably describes *Pseudocnus curatus* (Cowles). The following biology is from several papers that used the name *C. lubrica.* In the summer months, it feeds on particulate matter with its extended tentacles. No feeding occurs from October to March and females do not feed while brooding eggs. The diet is mostly single-celled algae.

Animals spawn from mid November to mid December around southern Vancouver Island. Males lift the anterior end off the substratum and release long strands of white sperm from genital papillae located between two of the dorsal tentacles. Sperm sink to the substratum and become entangled in the bodies of other sea cucumbers. Females raise the anterior end of the body off the substratum and arch backwards. Eggs are then spawned and roll down onto the female's ventral surface between her tube-feet. The tentacles are not used to catch the eggs; thus, many can be washed away by the current. The female then re-attaches to the rock, holding the eggs against the substratum. The large eggs (mean 973 μm) are brooded from January to March. Six weeks after spawning young cucumbers hatch, but stay with the female for another four to eight weeks. In the photo of *Pseudocnus curatus,* note the small, white juveniles beneath the parent and in the background.

Many species of sea stars eat *P. curatus:* Sand Star (*Luidia foliolata*), three species of sunstars (*Solaster stimpsoni, S. dawsoni, S. endeca*), Leather Star (*Dermasterias imbricata*), Six-armed Star (*Leptasterias hexactis*) and the Sunflower Star (*Pycnopodia helianthoides*). The Saddleback amphipod (*Parapleustes*) has been observed eating eggs. Unlike some other sea cucumbers, *P. curatus* does not show an escape response to any of the sea stars.

Presumably as a camouflage, adult *P. curatus* often attach pieces of shell, wood and other material to its dorsal surface. *P. curatus'* body wall is toxic to certain fish such as gunnels, *Apodichthys* and *Pholis:* a strong defence mechanism against these predators.

The parasitic gastropod, *Thyonicola mortenseni*, infects *Pseudocnus curatus.*

References
Arndt et al. (1966), Atwood (1974a, 1975), Atwood and Chia (1974), Birkeland et al. (1982), Chia et al. (1975), Cowles (1907), Engstrom (1974, 1982), Lambert (in press), McEuen (1987, 1988), Stricker (1986).

Pseudocnus lubricus (H.L. Clark, 1901b)

Originally as:	*Cucumaria lubrica* H.L. Clark
Revised to:	*Pseudocnus lubricus* by Lambert (in press)
Common name:	Aggregating Sea Cucumber
Synonym:	*Cucumaria fisheri astigmata* Wells, 1924
lubrica =	slippery

Description

Pseudocnus lubricus is a small, white to brownish black sea cucumber, often with fine black specks on the dorsum. It can grow up to 4 cm long. It is bluntly pointed at each end when retracted. The dorsal side varies from almost black to yellowish white. The underside is usually lighter than the top. The tube feet on the ventrum are well developed in double rows, with some scattered between the rows. On the dorsum, the tube feet are scattered and appear as lighter coloured bumps or dimples. The 10 tentacles are equal in size and extend upwards at an angle from the body which is usually firmly attached to the substratum. See Photos 16 & 17.

Fig. 38. Live *Pseudocnus lubricus* collected near Sooke, B.C.

Skin ossicles: vary from thick, knobbed, four-holed buttons to larger knobbed plates, some are elongated and quite thick with a spiny handle at one end, often referred to as pine-cone ossicles; three-armed ossicles from tube feet.

Lambert (in press) has placed this species and *C. curata* into the genus *Pseudocnus* based on the distribution of tube feet, the complex button ossicles and DNA evidence.

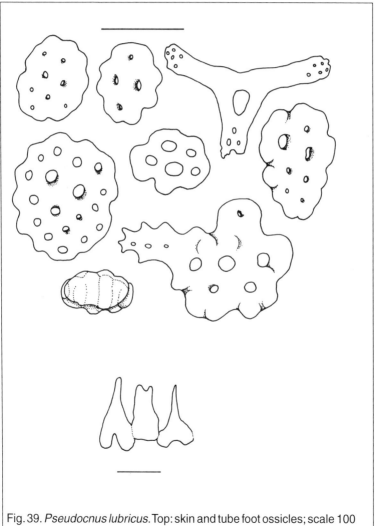

Fig. 39. *Pseudocnus lubricus.* Top: skin and tube foot ossicles; scale 100 μm. Bottom: dorsal part of calcareous ring; scale 2 mm.

Similar Species

In the literature, *Pseudocnus lubricus* has been confused with *C. curata, C. pseudocurata* and *C. piperata*. What used to appear in identification keys as *C. lubrica* — the small black aggregating species in the Juan de Fuca Strait region — is actually *Pseudocnus curatus* (=*C. curata*). True *P. lubricus (*=*C. lubrica)* is usually plain creamy white to brown, and often speckled with black. Consult the descriptions of the above species for comparative details. *Cucumaria fisheri astigmatus* Wells, 1924 also as *Pseudocnus astigmatus* (Wells, 1924) is considered to be a junior synonym of *C. lubrica* Clark (Arndt, Marquez, Lambert and Smith 1996), but *C. lubrica* has been revised to *P. lubricus* by Lambert (in press). The black spots of *C. piperata* are blotches rather than specks, and they occur all over the body rather than just on the dorsal surface. *C. piperata* normally buries itself in the substratum or under rocks.

Range

Recorded from southeastern Alaska to central California (this study); to Cortez Bank, southern California (Bergen 1996). Records of *Pseudocnus astigmatus* or *Cucumaria fisheri astigmatus* from California are synonymous with *P. lubricus*. Occurs from low inter-tidal down to 78 m on rocky substrata; most common on the exposed coast, but has been recorded as far as the San Juan Islands, eastern Juan de Fuca Strait. Because of the taxonomic confusion in the literature, the true range and habitat of this species are not yet clear.

Habitat

Pseudocnus lubricus is abundant subtidally on rock surfaces in large aggregations of yellowish white individuals. More common on the outer coast, but also found around the holdfasts of low intertidal and subtidal algae in Juan de Fuca Strait. Some brown specimens with black specks have been collected near Victoria, British Columbia, in the low intertidal zone. I have seen *P. lubricus* subtidally in large aggregations in the Broken Group Islands of Barkley Sound, and at the south end of Vancouver Island. We need more information to document its range of habitats.

Biology

In winter, *Pseudocnus lubricus* broods large, yolky eggs between its body and the substratum until the larvae develop and crawl out. Most of the literature on the biology of *C. lubrica* is probably referring to *C. curata* (now *Pseudocnus curatus*). These two species are closely related and may have similar reproductive behaviour, but due to past mistakes in identification this aspect of their biology is uncertain.

References

Arndt et al. (1996), Bakus (1974), Bergen (1996), Cherbonnier (1951), Clark (1901a, 1901b, 1902, 1924), Hetzel (1960), Lambert (1990b), Lambert (in press), Shick (1983), Tikasingh (1960), Wells (1924).

Ekmania diomedeae (Ohshima, 1915)

Originally as:	*Phyllophorus diomedeae* Ohshima
Revised to:	*Ekmania diomedeae* by Hansen and McKenzie (1991)
diomedeae =	Diomedes, a Trojan hero

Description

Ekmania diomedeae has a whitish or reddish sausage-shaped body that turns brownish in alcohol. Its maximum length is 15 cm. It has tube feet scattered uniformly over its body. The body wall tends to be thin and translucent. It has 15 tentacles: an outer circle of 10 larger tentacles (5 pairs), and an inner circle of 5 smaller single tentacles. See Photo 18.

Skin ossicles: not plentiful in the body wall, but do not deteriorate with age as in *Thyonidium kurilensis*. Table ossicles have a low spire with a crown of spines, a disk with scalloped margins and a single set of holes around the perimeter. In the introvert the ossicles are complex, oval perforated plates, some with bumps coalesced into a raised structure in the centre.

Fig. 40. Preserved *Ekmania diomedeae* from Kodiak Island, Alaska.

Similar Species

Ekmania diomedeae is similar to *Thyonidium kurilensis* and *Thyone benti.* The latter two also have scattered tube feet, but the number of tentacles separates them from *E. diomedeae. T. kurilensis* has 20 tentacles and *T. benti* has 10. *Ekmania barthii*, known from northeastern North America across the North Atlantic to Scandinavia, is very similar to *E. diomedeae,* and probably closely related.

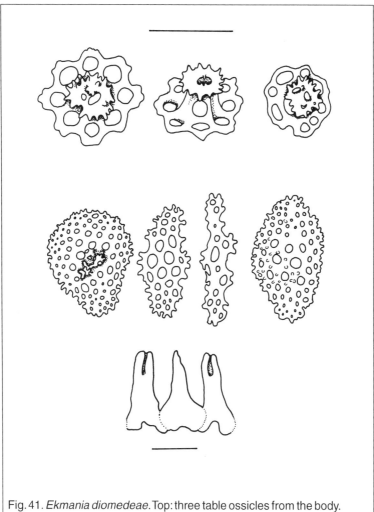

Fig. 41. *Ekmania diomedeae.* Top: three table ossicles from the body. Middle: four perforated plates from introvert. Bottom: dorsal part of calcareous ring. Top scale 100 μm, bottom scale 2 mm.

Range

Not well documented in the north Pacific since it was originally described from northern Japan. We have records from Forrester Island, off southeastern Alaska, near Kodiak Island in the Gulf of Alaska, and in the Chukchi Sea. From 37 to 220 m depth.

Habitat

Ekmania diomedeae is dredged from mud or sand-gravel substrata.

Biology

The biology of this species is unknown.

References

Hansen and McKenzie (1991), Heding (1942), Lambert (1984) (as *Thyonidium pellucidum*), Ludwig (1886), Mortensen (1932, 1977), Ohshima (1915), Ostergren (1902).

Thyonidium kurilensis (Levin, 1984)

Originally as: *Duasmodactyla kurilensis* Levin, 1984
kurilensis = named for the type locality: Kuril Islands, Russia

Description

Thyonidium kurilensis is an uncommon species with a whitish to pale orange body and reddish-orange tentacles. It turns whitish to grey with a tinge of purple in alcohol. Its body is thick-skinned, and tapers bluntly toward the posterior. Its maximum length is 15 to 20 cm. Its tentacles and mouth region are often a dark purple. *Thyonidium kurilensis* has 20 tentacles arranged in an inner circle of 5 small pairs and an outer circle of 5 larger pairs. It has tube feet scattered over the body surface, but rows can also be discerned among them. In juveniles the tube feet are in more obvious rows. See Photo 19.

Skin ossicles: tables with scalloped margins, numerous holes and a tall narrow spire, as well as large plates without a spire (usually only in the introvert). With age, ossicles become scarce in the body wall except for large circular end plates (200-300 μm in diameter) from the tube feet.

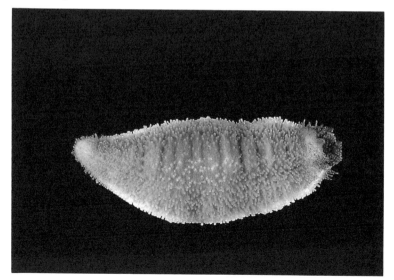

Fig. 42. Live *Thyonidium kurilensis* from San Juan Islands, Washington.

Similar Species

Externally, *Thyonidium kurilensis* is similar to *Ekmania diomedeae* and *Thyone benti*, but the number of tentacles distinguish them. *Ekmania diomedeae* has 15 tentacles, an outer circle of 10 larger tentacles (5 pairs) and an inner circle of 5 smaller tentacles alternating with the large pairs. *Thyone benti* has 10 tentacles, with 2 of them reduced in size.

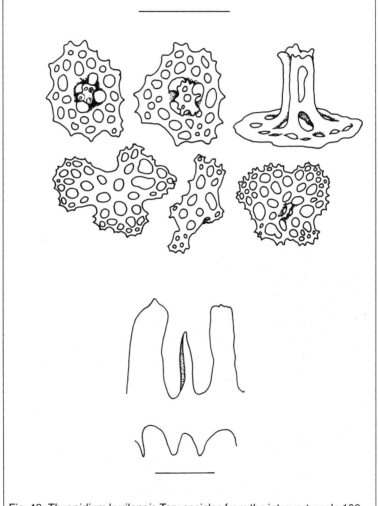

Fig. 43. *Thyonidium kurilensis.* Top: ossicles from the introvert; scale 100 µm. Bottom: dorsal part of calcareous ring; scale 5 mm.

Range

Two live specimens were collected from shallow water near Spieden Island in the San Juan Islands of Puget Sound. Other preserved specimens were from the Gulf of Alaska, Malcolm Island in Queen Charlotte Strait and Cordero Channel. Prior to these four records from the northeastern Pacific, *Thyonidium kurilensis* was known only from the Kuril Islands near Kamchatka, Russia. I would expect to find it in the Bering Sea, but so far there is a gap in the distribution between Kamchatka and Kodiak Island. Our records indicate a depth of 10 to 228 m in the north Pacific. It is similar to *Thyonidium drummondii*, a common species in the northern British Isles, Scandinavia, Iceland, Greenland and also reported from New England. Although it only differs from the Atlantic species in some minor ways, *T. kurilensis* should be considered a separate species unless at some later date the species is shown to be distributed across the Canadian Arctic.

Habitat

Thyonidium kurilensis is dredged from soft sediments. Andy Lamb of the Vancouver Public Aquarium collected two specimens from a gravelly substratum while he was scuba diving near Spieden Island in the San Juan Islands.

Biology

So far, we know nothing about the biology of the species on the west coast. Even the common Atlantic species, *Thyonidium drummondii*, is poorly studied.

References

Forbes (1841), Heding (1942), Heding and Panning (1954), Levin (1984), McKenzie (1991), Mortensen (1977), Pawson (1977), Théel (1886), Thompson (1840).

Family Phyllophoridae

Subfamily Thyoninae

Pentamera lissoplaca (Clark, 1924)

Pentamera populifera (Stimpson, 1864)

Pentamera pseudocalcigera Deichmann, 1938b

Pentamera trachyplaca (Clark, 1924)

Pentamera sp. A

Pentamera sp. B

Thyone benti Deichmann, 1937

External Features: Body U-shaped, or cucumberlike. Body wall soft and pliable, or stiff. Tube feet in five series, or scattered all over body. Tentacles 10-25. Tentacles dendritic, equal in size, or eight large, two small.

Internal Features: Tentacle ampullae absent. Retractor muscles present. Respiratory trees Y-shaped. Rete mirable absent. Gonad double tuft. One madreporic body, attached to dorsal mesentery. Cuvierian organs absent.

Calcareous parts: Calcareous ring with long posterior projections. Mosaic of small pieces. Typical skin ossicles perforated plates, or tables.

Genus *Pentamera*

Six species of *Pentamera* are described on the following pages, but two of them have yet to receive a formal scientific name. My preliminary analysis suggest that they are significantly different from the known species of *Pentamera*.

Rather than delay the publication of this book, I chose to publish these preliminary descriptions pending publication in a formal scientific journal. This will alert people to be on the lookout for these other forms of *Pentamera*.

Pentamera lissoplaca (H.L. Clark, 1924)

Originally as: *Cucumaria lissoplaca* Clark
Revised to: *Pentamera lissoplaca* by Deichmann (1938b)
lisso placa =   smooth plates

Description

Pentamera lissoplaca is a small, white species up to 3.5 cm long. Its body may be straight or curved, and tapers to the posterior end, at least in relaxed animals. The long, non-retractile tube feet are in five distinct, crowded double rows and are slightly straw-coloured. The skin between the rows is smooth or wrinkled. It has ten smallish tentacles, including two much smaller ventral ones, that are usually brown or darker than the body. See Photo 20.

Skin ossicles: mostly larger, elongate diamond-shaped plates, and a less-common delicate table with a small spire; also curved supporting rods in tube feet.

Fig. 44. Preserved *Pentamera lissoplaca* from Pearse Island, B.C.

Similar Species

All the *Pentamera* are similar, and difficult to tell apart. To be sure, you will have to check the ossicles and compare them with other species in this book. As with other *Pentamera* species, *Pentamera lissoplaca* can also be confused with *Eupentacta quinquesemita* and *Eupentacta pseudoquinquesemita*. The skin ossicles of these latter two are distinct.

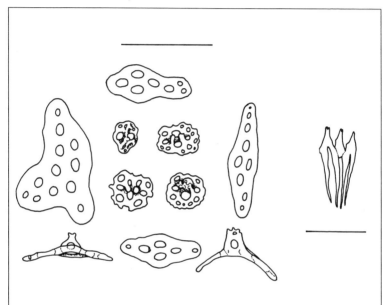

Fig. 45. *Pentamera lissoplaca*. Left: skin ossicles; scale 100 μm. Right: dorsal part of calcareous ring; scale 5 mm.

Range

The type locality is Alert Bay in Queen Charlotte Strait. Deichmann (1937) records it from southern Alaska south to Monterey Bay and a single specimen from Cedros Island, Baja California. I have examined specimens from Auke Bay and Wrangel, Southeast Alaska, and 10 localities in British Columbia from Portland Inlet, Queen Charlotte Islands, Barkley Sound and near Victoria. South to Magdalena Bay, Baja California (Bergen 1996). Intertidal down to 90 m.

COLOUR PHOTOGRAPHS

Each species pictured in this colour section has a
corresponding description in the body of the text. Each caption
in this section refers to the page number where the species
description begins.

Philip Lambert diving at Canoe Rock, B.C.

Photo 1.
Typical colour phase of *Parastichopus californicus,* Saanich Inlet, B.C. See species description on page 32.

Photo 2.
Parastichopus leukothele, Tasu Sound, Queen Charlotte Islands, B.C. See species description on page 36.

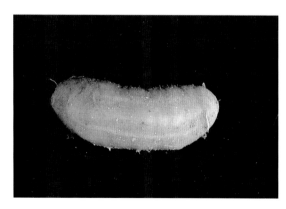

Photo 3.
Freshly dredged specimen of *Pseudostichopus mollis,* Queen Charlotte Sound. See species description on page 40.

Photo 4. Anterior end of *Synallactes challengeri,* Auke Bay, Alaska. See species description on page 42.

Photo 5. *Psolidium bidiscum,* Saanich Inlet, B.C. See species description on page 45.

3

Photo 6.
Psolus chitonoides,
McCurdy Point,
Saanich Inlet, B.C.
See species
description on page
48.

Photo 7.
A preserved
specimen of *Psolus
squamatus,* dredged
from Juan de Fuca
Canyon off the
southwest coast of
Vancouver Island.
See species
description on page
51.

Photo 8.
*Cucumaria frondosa
japonica* in a display
tank at Auke Bay
Marine Lab, Alaska.
See species
description on page
54.

4

Photo 9. Feeding tentacles of *Cucumaria miniata,* Ogden Point Breakwater, Victoria, B.C. See species description on page 58.

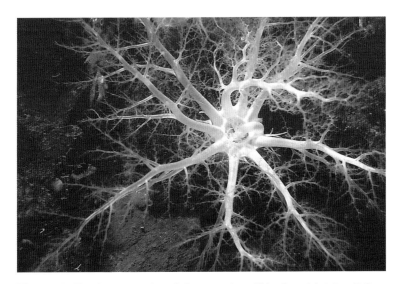

Photo 10. Feeding tentacles of *Cucumaria pallida,* Saanich Inlet, B.C. Note ten equal sized tentacles. See species description on page 61.

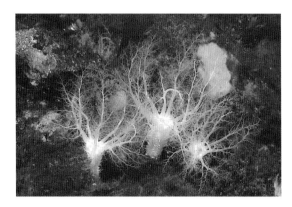

Photo 11.
Group of *Cucumaria pallida,* Saanich Inlet, B.C. See species description on page 61.

Photo 12.
Feeding tentacles of *Cucumaria piperata,* McCurdy Point, Saanich Inlet, B.C. See species description on page 64.

Photo 13.
Preserved *Cucumaria pseudocurata.* In life they are usually hidden among California mussels. See species description on page 67.

6

Photo 14. In the intertidal zone near Auke Bay, Alaska, black *Cucumaria vegae* along with white *Eupentacta pseudoquinquesemita* cover the rocks. See species descriptions on pages 71 and 110.

Photo 15. *Pseudocnus curatus* on shallow subtidal rock, Esquimalt, B.C. Note small white juveniles in the background and just beneath the parent. See species description on page 74.

7

Photo 16. *Pseudocnus lubricus* in a subtidal aggregation of thousands of individuals, Sooke, B.C. See species description on page 78.

Photo 17. A mottled brown *Pseudocnus lubricus* in the intertidal zone, Esquimalt, B.C. See species description on page 78.

Photo 18. A preserved specimen of *Ekmania diomedeae* dredged from a depth of 179 metres, east of Kodiak Island. See species description on page 82.

Photo 19. Two colour phases of *Thyonidium kurilensis* collected by Andy Lamb while scuba diving in the San Juan Islands. See species description on page 85.

Photo 20. A preserved specimen of *Pentamera lissoplaca* collected with aid of scuba from soft sediment in Winter Inlet, Pearse Island, B.C. See species description on page 89.

Photo 21.
A live *Pentamera populifera* dredged from 60 metres, Bosun Bank, Satellite Channel, B.C. See species description on page 92.

Photo 22.
A live *Pentamera pseudocalcigera* dredged from 60 metres, Bosun Bank, Satellite Channel, B.C. See species description on page 95.

Photo 23.
A preserved specimen of *Pentamera trachyplaca,* Houston Stewart Channel, Queen Charlotte Islands. See species description on page 98.

Photo 24. A preserved *Pentamera sp. A* dredged off Estevan Point, B.C. See species description on page 101.

Photo 25. A preserved *Pentamera sp. B* collected with the aid of scuba at Quatsino Sound, B.C. Note tuft of gonad protruding through damaged body wall. See species description on page 104.

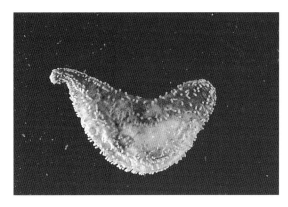

Photo 26.
A preserved specimen of *Thyone benti* dredged from 125 metres, Fitz Hugh Sound, B.C. See species description on page 106.

Photo 27.
Eupentacta pseudo-quinquesemita collected with the aid of scuba, Ogden Point Breakwater, Victoria. See species description on page 110.

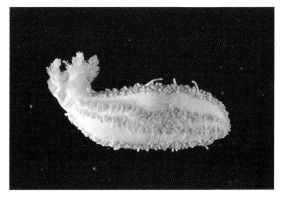

Photo 28.
Eupentacta quinquesemita in strong current near Victoria, B.C. See species description on page 113.

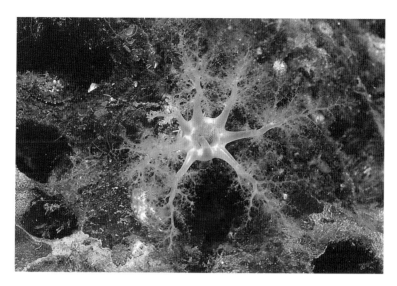

Photo 29. Feeding tentacles of *Eupentacta quinquesemita*, Colburne Passage, B.C. Note that it has eight large tentacles and two much smaller ones. See species description on page 113.

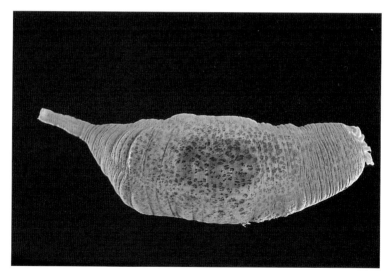

Photo 30. Preserved specimen of *Molpadia intermedia* dredged in Barkley Sound. See species description on page 118.

13

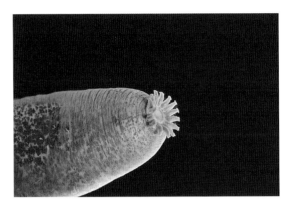

Photo 31.
Feeding tentacles of
Molpadia intermedia.
See species
description on page
118.

Photo 32.
A preserved
specimen of
*Paracaudina
chilensis* collected
from 11 metres near
Tow Hill, Queen
Charlotte Islands.
See species
description on page
122.

Photo 33.
A live *Leptosynapta
clarki* from Whiffin
Spit near Sooke, B.C.
See species
description on page
126.

Photo 34. A group of *Chiridota albatrossii* excavated from a shell substrate in Saanich Inlet, B.C. See species description on page 135.

Photo 35. *Chiridota discolor* found under a rock in the intertidal zone near Auke Bay, Alaska. See species description on page 135.

Photo 36. The white tentacles of *Cucumaria pallida* and the arms of brittle stars contrast with the orange tentacles of *Cucumaria miniata* at the Victoria breakwater. See species descriptions on pages 61 and 58.

Photo 37. A rich green soup of plankton bathes the feeding tentacles of *Cucumaria miniata* on a current-swept reef in the Gulf Islands of British Columbia. See species description on page 58.

Habitat

Pentamera lissoplaca is usually dredged from sand or soft mud. I have also collected it with aid of scuba in less than 10 metres from soft mud in northern British Columbia.

Biology

Unknown.

References

H.L. Clark (1924), Deichmann (1937, 1938b), Lambert (1984).

Pentamera populifera (Stimpson, 1864)

Originally as:	*Pentacta populifer* Stimpson
Revised to:	*Pentamera populifera* by Deichmann (1938b)
populifera =	**abundant**

Description

Pentamera populifera is short (2-3 cm) and stubby, pale yellowish white, and has five series of slender tube feet. The body is usually U-shaped or curved and often has a nipplelike tail. Eight of the ten tentacles are equal in size, two are one-tenth the size of the others. The thin, flexible skin between the rows of tube feet bristles with ossicles. See Photo 21.

Skin ossicles: tables with a two-pillared spire and an oval or star-shaped disk; a few plain perforated plates; curved supporting tables with a moderately tall spire.

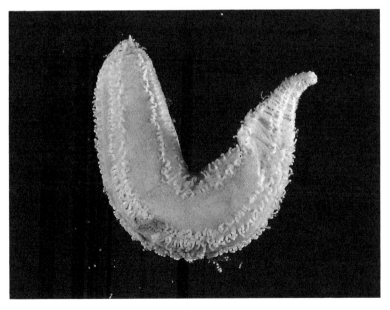

Fig. 46. Preserved *Pentamera populifera* from Saanich Inlet, B.C.

Similar Species

Pentamera populifera could be confused with *P. pseudocalcigera*, *P.trachyplaca* or *P. lissoplaca*. The differences are subtle, and one needs to analyse the ossicles to be certain. The first two mentioned are the most common. The tube feet differ in appearance, with *P. pseudocalcigera* having conical tube feet as opposed to cylindrical. Both tend to have a U-shaped body. The few specimens of *P. trachyplaca* that I have seen tend to be straighter and not as stiff. Two unnamed species of *Pentamera* that have previously been included in the *P. populifera* complex, are included at the end of this *Pentamera* section.

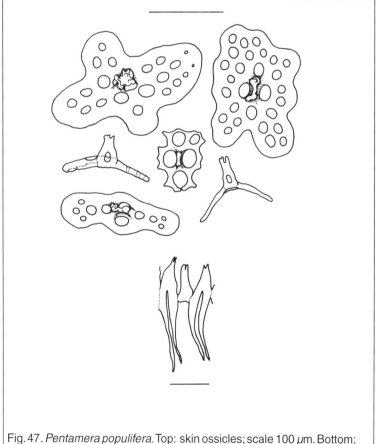

Fig. 47. *Pentamera populifera.* Top: skin ossicles; scale 100 μm. Bottom: dorsal part of the calcareous ring; scale 1 mm.

Range

Gulf of Alaska to Cedros Island, Baja California; shallow subtidal to 256 m.

Habitat

Large numbers of *P. populifera* inhabit the sediments of shallow, protected bays in Puget Sound and San Juan Islands. Up to 1400 specimens per square metre can live in the top 5-7 cm. Some dwell in sand near Cortes Island, Georgia Strait. They form a U-shaped burrow with each end at the surface.

Biology

Pentamera populifera is both a deposit and a suspension feeder. In captivity, young begin feeding on diatoms when only 17 days old. Adults spawn from mid February to late March. The green ovary, visible through the thin body wall, produces a light green egg (mean diameter 372 μm). The female entangles the eggs in her tentacles and lifts them up into the current. The testis is creamy orange. In culture the lecithotrophic larva (Fig. 8B) is attracted by light. The Sunflower Star (*Pycnopodia helianthoides*) is a major predator.

References

Clark, H. L. (1924), Deichmann (1937, 1938b), Lambert (1984), McEuen (1987, 1988).

Pentamera pseudocalcigera Deichmann, 1938b

pseudocalcigera = **false calcigera**

Description

Pentamera pseudocalcigera is U-shaped, and can grow up to 10 cm long. Its anterior end is bluntly pointed; the posterior end is longer and more tapered. This species is usually white or beige. In preserved specimens the tube feet are yellowish white, and on a slightly raised ridge above the smooth, pale purple skin. The bands of tube feet are made up of two to four rows each. Each tube foot has a broad base that tapers quickly to the tip. The introvert and 10 tentacles usually have peppery brown spots. The two ventral tentacles are about half the size of the others. See Photo 22.

Skin ossicles: large triangular to star-shaped perforated plates; a few may have a minute spire in the centre; curved supporting rods in the tube feet with a large complex spire.

Fig. 48. Live *Pentamera pseudocalcigera* from Dundas Island, B.C.

Similar Species

Compared to other *Pentamera* species, *P. pseudocalcigera* is larger. It has a long tapering tail rather than a nipplelike tail; and it has conical, rather than thin, cylindrical tube feet.

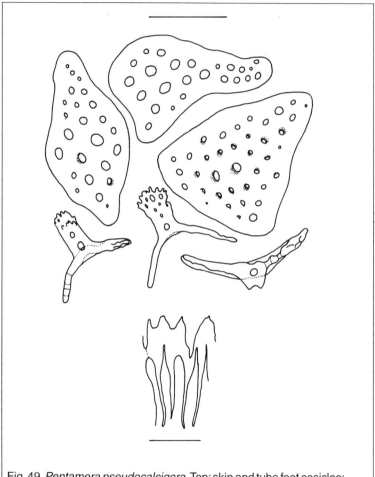

Fig. 49. *Pentamera pseudocalcigera.* Top: skin and tube foot ossicles; scale 100 μm. Bottom: dorsal part of the calcareous ring; scale 5 mm.

Range

Southeastern Alaska to southern California; 22 to 300 m. South to San Benitos Island, Baja California; 13 to 448 m (Bergen 1996).

Habitat

P. pseudocalcigera is commonly dredged from mud or sandy mud in well-flushed areas — not within fjords that have limited circulation. Its U-shape implies that the two ends lie at the mud surface with the body buried below.

Biology

Little is known about the biology of this species. It most likely feeds on organic deposits in the mud.

References

Alton (1972), Bergen (1996), Carney and Carey (1976), Deichmann (1938b), Edwards (1907), Lambert (1984, 1990a).

Pentamera trachyplaca (H.L. Clark, 1924)

Originally as:	*Cucumaria trachyplaca* Clark
Revised to:	*Pentamera trachplaca* by Deichmann (1938b)
Synonym:	*Cucumaria cosmotyrsitus* Wells, 1924
trachyplaca =	rough plate (referring to the ridged, knobbed skin ossicles)

Description

Pentamera trachyplaca is a small species that can grow up to 3 cm long. Its body is straight, and only slightly tapered at the ends. It is whitish or flesh coloured. Tube feet are in five crowded bands, with about four rows of tube feet in each band. The tube feet are cylindrical and bristle with ossicles. The bands of tube feet are about equal in width to the intervening space. It has ten darker tentacles, of which the two ventral ones are the smallest. See Photo 23.

Skin ossicles: plates with a complex knobbed surface on one side, often built up into a spire; in side view the plate is curved with the knobbed surface on the convex side.

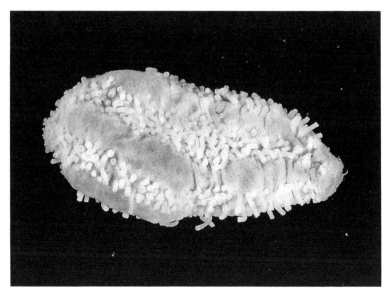

Fig. 50. Preserved *P. trachyplaca* from Houston Stewart Channel, B.C.

Similar Species

Pentamera trachyplaca can be confused with: *P. populifera*, *P. lissoplaca*, *P. pseudocalcigera*, *Eupentacta quinquesemita* and *E. pseudoquinquesemita*. The *Pentamera* species are similar externally, so ossicles should be checked to be certain. The *Eupentacta* species are usually straight rather than U-shaped, and the tube feet give them a spiky look. The external differences among all the above species are subtle, so ossicles should be examined.

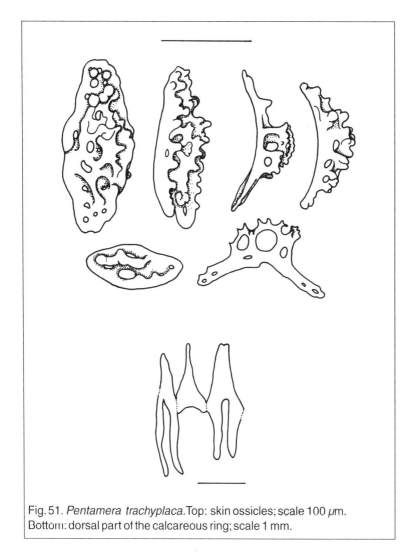

Fig. 51. *Pentamera trachyplaca*. Top: skin ossicles; scale 100 μm. Bottom: dorsal part of the calcareous ring; scale 1 mm.

Range

Known from only a handful of localities. From Banks Island on the northern coast of British Columbia; Cape St. James, Queen Charlotte Islands; Alert Bay, Queen Charlotte Strait; Barkley Sound, west coast of Vancouver Island; Monterey, California as *C. cosmotyrsitus* and Santa Cruz Island (Bergen 1996). Intertidal to 27 m.

Habitat

Pentamera trachyplaca is found in sand and mud in cobble areas.

Biology

Unknown.

References

H.L. Clark (1924), Deichmann (1938b), Lambert (1984), Wells (1924).

Pentamera sp. A

Probably an undescribed species

Description
This white species is up to 9 cm long and about 1 cm thick at its midpoint. The body is usually U-shaped, fat in the middle and long and tapered at each end. The skin is stiff with ossicles. The non-rctractile tube feet are in five distict bands, with two to four rows in each. The tube feet tend to be reduced in size and number toward the anterior and posterior ends of the body. Two of the ten tentacles are reduced in size. The tentacles and mouth area are speckled brown. See Photo 24.

Skin ossicles: round to triangular perforated plates (100-300 μm) with a large central spire covering half the width of the plate; curved supporting tables in the tube feet have a low bumpy spire. Tentacle ossicles: in two forms; finer, curved, elongate, oval plates and large, more robust rods with a few holes.

Fig. 52. Preserved *Pentamera sp. A* from Satellite Channel, B.C.

Similar Species

Pentamera sp. A is similar to *Pentamera populifera* and usually combined with it in collections. *Pentamera sp. A* is more elongate, thinner and stiffer. The table ossicles have a low, broad central spire and the supporting rods have a low spire. In contrast *P. populifera* has skin ossicles with a tall, narrow central spire and supporting rods with a tall spire. Most previous references have lumped this undescribed species in with either *P. populifera* or *P. pseudopopulifera* found in southern California. *Pentamera sp. A* is also similar in general appearance to *Pentamera sp. B,* but the skin ossicles are quite different.

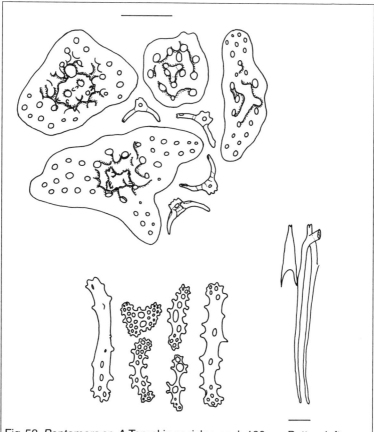

Fig. 53. *Pentamera sp. A.* Top: skin ossicles; scale 100 μm. Bottom left: tentacle ossicles. Bottom right: dorsal part of calcareous ring; scale 1 mm.

Range

Porcher Island, northern British Columbia south to the Strait of Georgia; on the outer and inner coast; 18 to 421 metres.

Habitat

Pentamera sp. A is usually dredged from sand or gravel substrata.

Biology

Nothing is known about its biology. Judging by its shape, I would speculate that it has its anterior and posterior ends at the surface of the mud.

References

Bergen (1996). I suspect that some specimens identified as *P. populifera* in this publication are *Pentamera sp. A*; Lambert (in preparation).

Pentamera sp. B

Probably an undescribed species

Description
This small, white to beige species grows up to 7 cm long. Its curved body is tapered at the ends, but is not as stiff as *Pentamera sp. A*. It has numerous fine tube feet in five bands, reduced in number at the extremities. The ten tentacles consist of eight large and two small, ventral tentacles. See Photo 25.

Skin ossicles: small roundish tables with four main holes and four smaller holes and a two-pillared spire; the supporting tables are slightly curved with a low spire; the introvert region has similar ossicles, but a bit smaller with fine teeth on the spire. Tentacles have two types of osssicles; finer, oblong, curved, perforated plates with two larger central holes; and large robust curved rods with a few holes.

Similar Species
Pentamera sp. B could be confused with *Pentamera sp. A* and *Pentamera populifera*. The photo illustrations show the differences in general appearance, but one should check the skin ossicles to be sure. Sometimes differences in shape may be due to how the specimens were preserved.

Range
Fitz Hugh Sound on the central coast of British Columbia south to Cape Flattery; 7 to 150 metres.

Habitat
Pentamera sp. B has been collected from a variety of substrata — mud, gravel, sand and cobble.

Biology
The biology of this new species is unknown.

Reference
Lambert (in preparation).

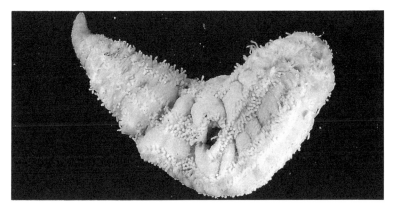

Fig. 54. Preserved *Pentamera sp. B* from Quatsino Sound, B.C.

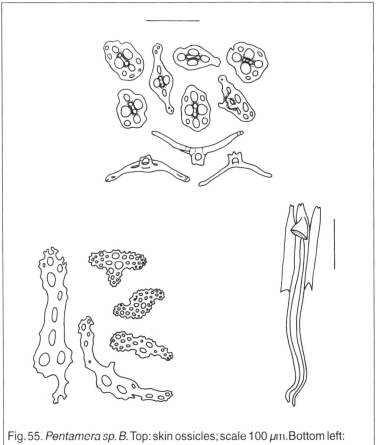

Fig. 55. *Pentamera sp. B.* Top: skin ossicles; scale 100 μm. Bottom left: tentacle ossicles. Bottom right: dorsal part of calcareous ring; scale 5 mm.

Thyone benti Deichmann, 1937

Revised to: *Havelockia benti* by Panning (1949), but in a postscript in the same paper he returned the species to genus *Thyone.*

benti = possibly referring to curve of body

Description

Thyone benti is small to medium-sized (5 to 15 cm long). It tapers to a pointed posterior end, and is usually curved. Preserved specimens are usually blackish brown at the extremities and lighter in the middle. I have not seen a live specimen. Tube feet are in rows at the posterior and anterior, but also scattered between. The 10 tentacles are light brown or orange, short and stubby, with the two ventral ones much smaller. See Photo 26.

Skin ossicles: round, oval or diamond-shaped tables with or without a two-pillared spire; curved supporting rods with a low spire. The number of ossicles in the skin declines with age. The introvert has mostly tables with more complex discs than in the body (not shown).

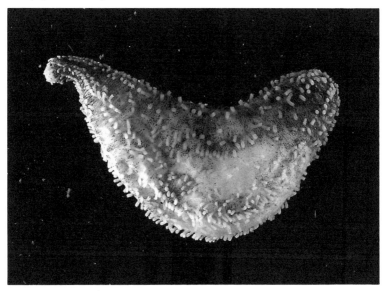

Fig. 56. Preserved *Thyone benti* from Fitz Hugh Sound, B.C.

Similar Species

Thyone benti is similar to *Ekmania diomedeae* and *Thyonidium kurilensis,* with its scattered tube feet, but the number of tentacles and the form of the skin ossicles will separate them.

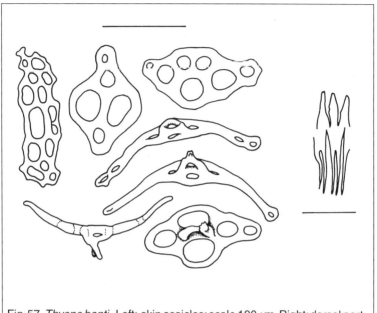

Fig. 57. *Thyone benti.* Left: skin ossicles; scale 100 μm. Right: dorsal part of the calcareous ring; scale 5 mm.

Range

Queen Charlotte Islands to Baja California; shallow subtidal to 135 m.

Habitat

Thyone benti is found in firm mud, sand or gravel. It is not common in B.C. waters. Specimens have been collected in the San Juan and Gulf Islands, Nanoose Bay, Malcolm Island and Cape Edensaw, usually with the aid of a dredge. Few other specimens have been recorded.

Biology

According to McEuen (1987) *T. benti* is a nocturnal suspension feeder. Spawning may take place in winter, because a specimen collected in March appeared to have a spent ovary with relict eggs of 512-560 μm diameter. This size suggests a pelagic lecithotrophic larva. Little else is known about this species. It is not clear if the above observations are attributed to the correct species.

References
Bakus (1974), Deichmann (1937, 1938a), McEuen (1987), Mottet (1976).

Family Sclerodactylidae

Eupentacta pseudoquinquesemita Deichmann, 1938b
Eupentacta quinquesemita (Selenka, 1867)

External Features: Body cucumberlike, or U-shaped. Body wall soft and pliable, or stiff. Tube feet scattered all over body, or in 5 series. Tentacles 10-20. Tentacles dendritic, equal in size, or 8 large, 2 small. Internal Features: Tentacle ampullae absent. Retractor muscles present. Respiratory trees Y-shaped. Rete mirable absent. Gonad double tuft. One madreporic body, attached to dorsal mesentery. Cuvierian organs absent.

Calcareous parts: Calcareous ring with paired or unpaired, short to moderate posterior projections; not a mosaic of small pieces. Typical skin ossicles large, perforated, ovoid bodies, smaller ossicles baskets, or cups.

Eupentacta pseudoquinquesemita Deichmann, 1938b

pseudo	=	**false**
quinquesemita	=	**five foot paths**

Description

Eupentacta pseudoquinquesemita has the same body form as *E. quinquesemita,* with five rows of non-retractile tube feet and smooth spaces between. It grows up to 10 cm long. The skin is soft and pliable. The general colour is a creamy white with the tentacles being a faint peachy colour. Typically, bits and pieces of shell and other debris adhere to the tube feet. There are eight equal-sized tentacles and two tiny ones on the ventral side. The tentacles are usually retracted when the species is collected. See Photo 27.

Skin ossicles: numerous, porous ovoid bodies as well as more delicate cups. Cups are small oval plates shaped like a shallow dish with a rim of knobs. This species has few, if any, baskets.

Fig. 58. *Eupentacta pseudoquinquesemita* off Esquimalt, B.C.

Similar Species

The external differences between the two *Eupentacta* species are subtle and not always consistent. Externally, the tube feet of *E. pseudoquinqesemita* are finer and more numerous, with a broad space between the series of tube feet. It also has a softer body. Without having the two species side by side, this description may not help too much. To be sure, check the skin ossicles for the presence of baskets or cups, as described above. The habitat will provide a clue as to which species you have (see below). In British Columbia the most common intertidal species is *E. quinquesemita*, but in southeast Alaska *E. pseudoquinquesemita* is more common.

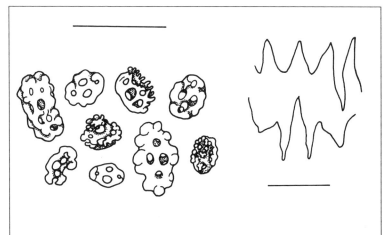

Fig. 59. *Eupentacta pseudoquinquesemita.* Left: skin ossicles; scale 100 μm. Right: dorsal part of the calcareous ring; scale 5 mm.

Range

Aleutian Islands to Puget Sound; intertidal to 200 m.

Habitat

In the southern part of its range, *E. pseudoquinquesemita* tends to be subtidal. It is often found on soft substrata. It commonly has particles of shell or seaweed attached to it. One study described it as not occurring on float communities, but restricted to exposed hard surfaces subject to currents (density as high as 100/m^2).

Biology

Little is known about the biology of this species, because it is so easy to confuse the two *Eupentacta*. Reports in the literature could refer to either species. Gut contents revealed a mixture of filamentous algae and diatoms as well as detritus and inorganic particles. In the field they were not seen to feed during the winter. Thirty-eight per cent of *Eupentacta pseudoquinquesemita* specimens examined in Puget Sound contained parasites, primarily the gastropod *Thyonicola americana*. The wormlike, coiled parasite attaches to the posterior third of the intestine.

References

Deichmann (1938b), Engstrom (1974), Tikasingh (1960, 1961).

Eupentacta quinquesemita (Selenka, 1867)

Common names: White Sea Cucumber, White Gherkin
Originally as: *Cucumaria quinquesemita* Selenka
Revised to: *Eupentacta quinquesemita* by Deichmann (1938b)
quinquesemita = five foot paths

Description

Eupentacta quinquesemita is stiff to the touch due to abundant calcareous ossicles in the skin and tube feet. The body grows to 5 to 10 cm in length. The non-retractile tube feet give it a spiny look. It has five rows of tube feet (four tube feet in width) with smooth skin between. The two ventral tentacles are smaller than the other eight. This character is useful for identifying this species when only the tentacles are visible. The expanded tentacles are creamy white with tinges of yellow or pink at the bases. See Photos 28 & 29.

Skin ossicles: numerous large, porous, ovoid bodies dominate the ossicles but among them are small, delicate baskets. These latter are important in differentiating this species from *Eupentacta pseudoquinquesemita.*

Fig. 60. *Eupentacta quinquesemita* from Arbutus Island, B.C.

Similar Species

Eupentacta quinquesemita is almost identical in external appearance to *E. pseudoqinqesemita*. The differences are subtle, and separation of these two species can only be determined with certainty by analysis of the skin ossicles. *E. pseudoquinqesemita* has softer, non-retractile tube feet, up to 10 across each row. In British Columbia and Washington this species is subtidal and often found in sediment. They often have small pieces of shell and other debris attached to their tube feet. If only the tentacles are visible, *E. quinquesemita* can also be confused with the white tentacles of *Cucumaria pallida*. However, *C. pallida* has 10 equal-sized tentacles, not 8 equal and 2 small. *Pentamera trachyplaca* and *P. lissoplaca* are also similar in general appearance but the ossicles are clearly distinct.

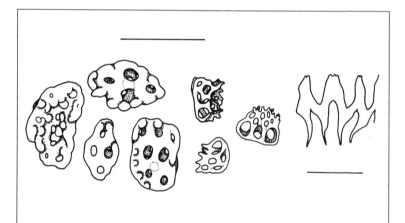

Fig. 61. *Eupentacta quinquesemita*. Left: skin ossicles; scale 100 μm. Right: dorsal part of the calcareous ring; scale 5 mm.

Range

Sitka, Alaska to Baja California; intertidal to 55 m.

Habitat

E. quinquesemita occurs along much of the coast in both protected and exposed regions. It is abundant in the intertidal and shallow subtidal of rocky shores. High densities of this species occur in strong currents. Juveniles (up to 1 cm) settle among hydroids and small algae in high current areas and on floating docks.

Biology

Eupentacta quinquesemita is a suspension feeder. It spawns from late March to mid May. The female produces light green eggs, 370 to 416 μm diameter: the male releases sperm, and fertilization takes place in open water. The yolky egg develops into a non-feeding evenly ciliated larva (Fig. 8A). In culture, the larva grows to the armoured stage in 11 to 16.5 days.

The Sun Star (*Solaster stimpsoni*), the Sunflower Star (*Pycnopodia helianthoides*), the Six-armed Star (*Leptasterias hexactis*) and the Kelp Greenling (*Hexagrammos decagrammus*) prey on this species.

Eupentacta quinquesemita occasionally eviscerates through a rupture in the introvert just behind the feeding tentacles. Rough handling causes this reaction, but it also happens under natural conditions and on a seasonal basis. Evisceration typically occurs from September to November and regeneration of the ejected parts takes about two to four weeks. Researchers believe that this behaviour is a method of discarding a waste-laden digestive tract, and also getting rid of parasites that attach to it.

The internal parasite, *Thyonicola americana*, a shell-less wormlike snail, attaches elongated coils of eggs to the intestine of *E. quinquesemita*. The larvae are released into the intestine and probably escape through the anus. Any parasites that are ejected by evisceration perish. For more details see Byrne (1985a).

The body wall of this species contains a poison that, if ingested, can cause a fish to become sluggish and eventually die.

References

Bakus (1974), Barr and Barr (1983), Bingham and Braithwaite (1986), Brumbaugh (1980), Byrne (1985a, 1985b, 1985c, 1986), Chia et al. (1975), H.L. Clark (1901a, 1901b), Engstrom (1974), Gotshall and Laurent (1979), Johnson and Johnson (1950), Mauzey et al. (1968), McEuen (1987, 1988), Mottet (1976), Sabourin and Stickle (1981), Shick (1983), Stickle and Denoux (1976), Stricker (1986), Tikasingh (1960).

Family Molpadiidae

Molpadia intermedia (Ludwig, 1894)

External Features: Body a smooth sausage with nipplelike posterior end. Body wall soft and pliable. Tube feet absent. Tentacles 15, digitate, equal in size.

Internal Features: Tentacle ampullae present. Retractor muscles absent. Respiratory trees Y-shaped. Rete mirable absent. One madreporic body. Cuvierian organs absent. Radial canals of water-vascular system present, but tube feet lacking.

Calcareous parts: Calcareous ring simple, not a mosaic. Typical skin ossicles are tables with a three-pillared spire, and perforated plates shaped like racquets. Some with red phosphatic bodies in skin.

Molpadia intermedia (Ludwig, 1894)

Common name: Sweet Potato Sea Cucumber
Originally as: *Trochostoma intermedium* Ludwig
Revised to: *Molpadia intermedia* by Clark (1907)
intermedia = may mean intermediate

Description

Molpadia intermedia has a smooth, sausage-shaped body that tapers abruptly to a short, appendix-like tail that is 20-25% of its body length. This species ranges in colour from a dark, purplish brown to purplish grey with rusty brown patches. Specimens collected with a dredge contract into plump sausages 11 to 14 cm in length. When relaxed and feeding, they can extend 35 to 43 cm in length. There are no tube feet on the smooth skin. The 15 tentacles — only 2 or 3 mm long — form a tight ring around the mouth. The stubby tentacles have a single pair of lateral branches just below the tip. See Photos 30 & 31.

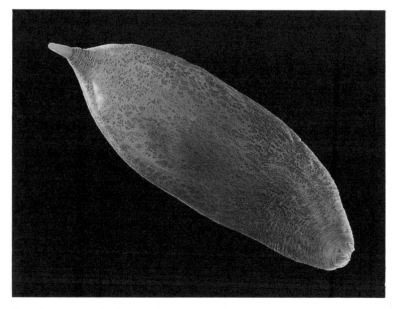

Fig. 62. Preserved *Molpadia intermedia,* dredged from the Strait of Georgia, B.C.

Skin ossicles: It is difficult to find ossicles in adult specimens. Reddish brown phosphatic bodies make up 99% of the deposits in the skin. Among the phosphatic bodies, and especially near the posterior end, there are a few variable tables, as illustrated in Figure 63. In juveniles there are clusters of racket-shaped plates radiating around a central table with a tall narrow spire (not shown).

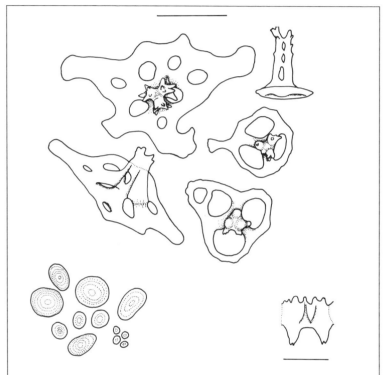

Fig. 63. *Molpadia intermedia*. Top: Table ossicles from tail. Bottom left: phosphatic bodies are the most common constituent. Bottom right: dorsal part of the calcareous ring. Top scale 100 µm; bottom scale 5 mm.

Similar Species
There are no species similar to *Molpadia intermedia* in the shallow waters covered in this book.

Range
Kodiak Island, Alaska to Gulf of Panama; 7 to 2,925 m.

Habitat

Molpadia intermedia is common in soft sediments in many coastal inlets. It is found up to 35 cm below the surface of the mud, but only 5 to 8 cm in more firm mud. This species is dominant in some areas where it reaches a density of up to 15 per m². Individuals tend to aggregate in groups of two to six in an area of about 1 to 2 m².

Biology

I have found no studies on the feeding habits of *Molpadia intermedia*. However, a similar species from the east coast, *Molpadia oolitica*, occurs vertically in the mud with its mouth pointing downward and its anus at the surface. Mud is eaten and forcibly expelled through the anus, forming a cone-shaped mound. *M. intermedia* may behave in a similar fashion.

Molpadia intermedia spawns from mid November to late December. Under study in a laboratory, females forcibly ejected negatively buoyant, orange-pink eggs of mean diameter at 267 μm from the gonopore at the anterior end. Males eject puffs of sperm over a period of 1.5 to 3 hours. The pelagic lecithotrophic larva (Fig. 8D) settles out of the plankton after about 96 hours. In the laboratory, they are negatively phototactic. Early juveniles feed in the top few millimetres of mud with the mouth facing up.

The skin of *M. intermedia* is characterized by ovoid, reddish-brown granules made of some calcium and silica, but mostly a form of iron called ferric phosphate: hence the name phosphatic bodies. The granule content increases with age, and large specimens from shallow water have more. Researchers are not certain of the function of these granules.

The Sand Star (*Luidia foliolata*) preys on *Molpadia intermedia*.

References

Alton (1972), Carney and Carey (1976), Clark, H. L. (1907, 1908), Deichmann (1937), Ellis (1969), Hetzel (1960), Lowenstam and Rossman (1975), Ludwig (1894), Mauzey et al. (1968), McEuen (1987, 1988), McEuen and Chia (1985), Nichols, F. H. (1975), Ohshima (1915), Stricker (1986), Terwilliger and Read (1970).

Family Caudinidae

Paracaudina chilensis (Müller, 1850)

External Features: Body long and tapered. Body wall soft and pliable. Tube feet absent. Tentacles 15, digitate, equal in size.

Internal Features: Tentacle ampullae present. Retractor muscles absent. Respiratory trees Y-shaped. Rete mirable present. Cuvierian organs absent. Radial canals of water-vascular system present, but tube feet lacking.

Calcareous parts: Calcareous ring with short posterior projections. Not a mosaic. Typical skin ossicles perforated plates, tables or cups.

Paracaudina chilensis (Müller, 1850)

Common names: Sand Sea Cucumber, Rattail Sea Cucumber
Originally as: *Molpadia chilensis* Müller
Revised to: *Paracaudina chilensis* by Heding (1931)
chilensis = first described from Chile

Description

Paracaudina chilensis has a smooth, elongated body that gradually tapers to a long narrow posterior end. It varies in colour from milky white to pinkish, purplish or silvery grey. This species measures up to 20 cm in length — the narrow tail making up to half the total length. The smooth, leathery skin has no tube feet. The 15 digitate tentacles have 2 pairs of short, lateral branches. See Photo 32.

Skin ossicles: vary from elaborate cups with wavy margins like a flower, to octagonal plates with a low central mound surrounded by small holes in older individuals.

Fig. 64. Preserved *Paracaudina chilensis* from Dundas Island, B.C.

Similar Species

There are no species in British Columbia waters that resemble *Paracaudina chilensis*. Its closest relative, *Molpadia intermedia*, has a nipplelike tail, brown or dark purple skin, and entirely different skin ossicles.

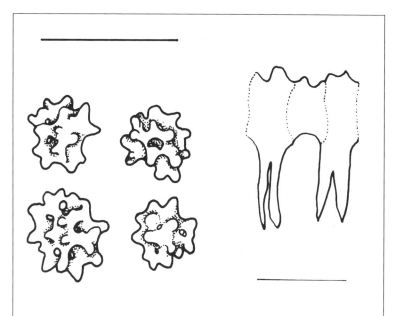

Fig. 65. *Paracaudina chilensis*. Left: skin ossicles; scale 100 μm. Right: dorsal part of the calcareous ring; scale 5 mm.

Range

Paracaudina chilensis is found all around the Pacific rim from the Straits of Magellan along coastal South, Central and North America to Japan, China, northern Australia and New Zealand. It has been recorded from 9 to 990 m depth throughout its range; but in British Columbia it is known only to 100 m.

Habitat

Paracaudina chilensis is a relatively uncommon species judging by the few specimens in museum collections — there are only three lots at the Royal British Columbia Museum. These were collected intertidally and in shallow subtidal areas on clean sand beaches off islands of northern B.C.

This species' burrow is marked by small conical mounds of sand that are ejected from its posterior end.

Biology

P. chilensis lives in sand. It ingests large quantities of sand, digests the organic particles and passes the sand out its anus. Scientists estimate that the animal can process 6 to 8 grams of sand per hour or 57 to 64 kg (125 to 140 lbs) annually. When placed on the sand, the anterior end always burrows down. The posterior end remains at or near the surface to expel sand and to obtain clean water for respiration. The cloaca expands and contracts to draw oxygenated water in, and on every fourth contraction, faeces are expelled.

This species spawns in May and June. The spawned eggs are yellow-brown, and about 500 μm. This species reaches full size in three or four years.

Paracaudina chilensis contains the respiratory pigment, haemoglobin.

No parasites or commensals are known from specimens in British Columbia. We might expect to find pea crabs associated with this species because Japanese specimens have the pea crab, *Pinnixa tumida*, in the cloaca.

References

Baker and Terwilliger (1993), Boolootian (1962), Carney and Carey (1976), Clark, H. L. (1907), Deichmann (1938a, 1947), Hozawa (1928), Inaba (1930), Kawamoto (1927), Kozloff (1987), Manwell (1959), Miller and Pawson (1984), Nomura (1926), Pawson (1963, 1965, 1970), Tao (1930), Yamanouchi (1926).

Family Synaptidae

Leptosynapta clarki Heding, 1928
Leptosynapta transgressor Heding, 1928

External Features: Body wormlike. Body wall soft and pliable. Tube feet absent. Tentacles 10-15, pinnate, equal in size.

Internal Features: Tentacle ampullae absent. Retractor muscles absent. Respiratory trees absent. Cuvierian organs absent. Radial canals of water-vascular system absent, only circumoral ring present.

Calcareous parts: Calcareous ring simple, not a mosaic. Typical skin ossicles: anchors and perforated anchor plates.

Leptosynapta clarki Heding, 1928

Common name:	**Burrowing Sea Cucumber**
Synonym:	*Leptosynapta roxtona* **Heding, 1928**
clarki =	**Dr H.L. Clark, an eminent American echinoderm specialist**

Description

Leptosynapta clarki is a wormlike sea cucumber. It averages 5 to 10 cm, but can reach 14.5 cm in length. The pale pinkish orange skin has numerous small papillae that may be tinged with orange, pink or red. Five longitudinal muscle bands show through the semi-transparent skin. The 12 (sometimes 10 to 14) short, clawlike pinnate tentacles usually have 5-8 pairs of lateral digits. This species has no tube feet or respiratory trees. Oxygen exchange occurs through the thin body wall. See Photo 33.

Skin ossicles: anchors associated with an oval perforated anchor plate; also simple, straight or curved rods in skin and tentacles. In *L. clarki* the anchors in the posterior part of the body are longer than the anterior body anchors.

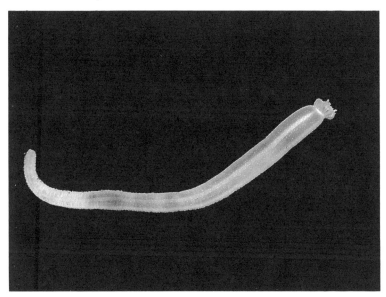

Fig. 66. Preserved *Leptosynapta clarki* from Barkley Sound, B.C.

Similar Species

Leptosynapta clarki is similar to *L. transgressor,* but *L. transgressor* averages only 10 tentacles, and its skin is white with dark red pigment spots and no papillae. *L. transgressor* is normally subtidal. In a recent paper, Sewell et al. (1995) suspect that *L. transgressor* is a subtidal ecological variant of *L. clarki.* However, until proven, I have retained it as a separate species. *L. roxtona* and *L. clarki* are also synonymous, according to Sewell et al.

Chiridota is another wormlike sea cucumber with reddish or purple skin and three irregular rows of small white bumps containing clusters of wheel ossicles.

Wormlike sea cucumbers can be distinguished from other wormlike organisms by 1) lack of segments 2) no body appendages or bristles (setae) 3) five longitudinal muscle bands 4) a tuft of short tentacles at one end 5) usually being smooth-skinned and 6) calcareous ossicles in the skin. Some burrowing anemones look similar, but have more than five internal septa that run the length of the body and show through the skin as white lines.

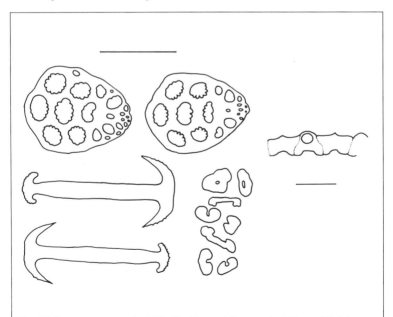

Fig. 67. *Leptosynapta clarki.* Left: skin ossicles; scale 100 μm. Right: dorsal part of calcareous ring; scale 1 mm.

Range
Queen Charlotte Islands to central California; primarily intertidal but reported to 73 metres. Some of the southern records may be a similar species, called *L. albicans*.

Habitat
Leptosynapta clarki usually inhabits mud flats among the roots of eelgrass (*Zostera*). This species tolerates a wide range of sediments from sandy silt to gravel. It occupies semi-permanent burrows 8 to 10 cm below the surface. It occurs in areas protected from direct waves on both inner and outer coasts of British Columbia and Washington.

Biology
Adult *L. clarki* ingest sediment and digest the organic content, just as earthworms do. Young *Leptosynapta* eat diatoms on eelgrass and other seaweeds. In eelgrass beds, the population of *L. clarki* may reach 245 per m².

Males shed sperm from mid November to mid December. Some biologists suspect that sperm enter the female via pores in the wall of the cloaca, or via the gonopore, to fertilize the eggs. The mean diameter of mature eggs recorded in a study at False Bay, San Juan Island, was 345 μm. The same study showed the maximum number of eggs was 2,495 per female. Females brood the eggs internally from November to April. During that time, yolk from nurse eggs nourish the developing juveniles. Juveniles are 12 mm long with 12 tentacles when they are expelled through circumanal ducts that connect the body cavity to the cloaca. This species is a protandric hermaphrodite: all individuals start out as males, and about half of those later become female. The sex change occurs when the individuals are between 200 and 400 mg in weight. There is some evidence that a sex change may occur more than once in its lifetime.

The microscopic, anchor-shaped ossicles (2 to 8 per mm²) form a bulge in the skin that enables the sea cucumber to grip the sediment while burrowing.

The Sand Star (*Luidia foliolata*), fish, sea gulls and crabs eat *Leptosynapta*.

The commensal polychaete worm, *Malmgrenia nigralba* (= *M. lunulata*), and the bivalve, *Scintillona bellerophon*, occur with *Leptosynapta clarki*. *S. bellerophon* attaches to the host with fine threads or with its foot in 57% of specimens collected at a Sooke Harbour site. Other species often associated with *Leptosynapta clarki* are the burrowing brittle star *(Amphiodia occidentalis)* the bivalves *(Macoma nasuta* and *Mysella tumida)* the Pea Crab *(Pinnixa schmitti)* and the polychaete worms *(Harmothoe lunulata* and *Pholoe minuta)*.

References
Anderson (1965, 1966), Atwood (1973, 1974a, 1974b, 1975), Berkeley (1924), Brenchley (1982), Brooks (1965), Chia and Burke (1978), Chia et al. (1975), Everingham (1961), Gibson and Burke (1983), Heding (1928), Hess et al. (1988a, 1988b), Hetzel (1960), Layton (1975), Mauzey et al. (1968), McDermid (1983), McEuen (1987, 1988), O'Foighil and Gibson (1984), Sewell (1994a, 1994b), Sewell and Chia (1994), Sewell et al. (1995), Stricker (1985, 1986), Taghon (1982).

Leptosynapta transgressor Heding, 1928

Common name: **Burrowing Sea Cucumber**
transgressor = **go beyond the limits (perhaps meaning of
other *Leptosynapta*)**

Description

The wormlike body of *Leptosynapta transgressor* has no tube feet.
It can grow up to 10 cm long. The body is white with fine, dark red
pigment spots on its smooth skin. The pigmentation is more dense
on the dorsal side and at the anterior end. The pinnate tentacles vary
from 9 to 14 with a mean of 10 (65% had 10). (No colour photo.)

Skin ossicles: anchors and anchor plates as with *L. clarki*; with
curved rods in skin and tentacles. The ossicle differences between
these species are subtle, such as the edges of tentacle rods are more
wavy in *L. transgressor*. We do not know whether these are real
species differences, or just due to ecological or habitat variability. In
fact, Sewell et al. (1995) suspect that *L. transgressor* is an ecological
variant of *L. clarki*. However, until such time as their taxonomic
status is revised, I have retained them as two separate species.

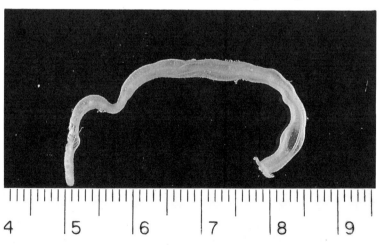

Fig. 68. Preserved *Leptosynapta transgressor* from Barkley Sound, B.C.

Similar Species

Leptosynapta transgressor can be easily confused with *L. clarki.* Both are quite variable depending on their age and size. The number of tentacles provides a clue; but there is a broad overlap in the number. The habitat and depth also help to differentiate the two species — with *L. transgressor* being subtidal and *L. clarki,* intertidal. In live specimens the colours differ. The ossicles are too similar to be useful in separating them. Only colour, texture and number of tentacles seem to differ. As noted for *L. clarki,* these two species may prove to be synonymous.

Chiridota albatrossii is another wormlike species, but it is pink or purple with white dots. The dots are small sacs filled with a cluster of wheel ossicles.

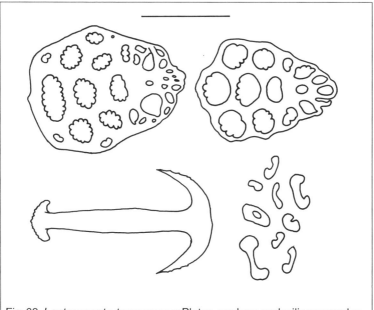

Fig. 69. *Leptosynapta transgressor.*Plates, anchors and miliary granules from the skin, scale 100 μm.

Range

Known distribution is limited to lower Vancouver Island and Puget Sound. It may prove to be more widespread if it could be easily differentiated from *L. clarki*. Until the relationship of these two species of *Leptosynapta* is worked out, distribution cannot be properly determined.

Habitat

Leptosynapta transgressor is a subtidal species that occurs in bays and channels of the inside waters between 6 and 40 m. It is commonly found in sandy silt and medium sand. The faunal community associated with *L. transgressor* is richer than with *L. clarki* (55 species vs 29), including polychaete worms, brittle stars, amphipods, bivalves and gastropods.

Biology

L. transgressor ingests sediment like other burrowing sea cucumbers. Its eggs reach their largest size in winter, and appear to be fertilized in April. There is evidence that the species broods its eggs, but only early embryological stages have been found in females. Little else is known about the biology of this species. The commensal polychaete, *Malmgrenia lunulata*, has been found with *Leptosynapta transgressor*.

References

Brooks (1965), Heding (1928, 1931), McEuen (1987), Sewell et al. (1995).

Family Chiridotidae

Chiridota albatrossii Edwards, 1907
Chiridota discolor Eschscholtz, 1829
Chiridota laevis (Fabricius, 1780)
Chiridota nanaimensis Heding, 1928

External Features: Body wormlike. Body wall soft and pliable. Tube feet absent. Tentacles 10-18, digitate, or pinnate, equal in size.

Internal Features: Tentacle ampullae absent. Retractor muscles absent. Respiratory trees absent. Rete mirable absent. Cuvierian organs absent.

Calcareous parts: Calcareous ring simple, not a mosaic. Typical skin ossicles: wheels or narrow rods.

Chiridota species

chirido (Gr.)   =   **sleeve**

Description

Several species of *Chiridota* have been described from the waters of the northeastern Pacific. The species are poorly defined and difficult to distinguish; so until the taxonomy is revised, I cannot present definitive descriptions. *C. albatrossii* has the most complete description and is probably the common deepwater species.

All species are wormlike, varying in length up to 30 cm. Five longitudinal muscles and other organs are visible through the translucent skin. A series of white bumps (wheel papillae) of varying number occur in rows along the length of the body. These papillae contain clusters of microscopic wheel ossicles. The top pair in Figure 71 is from a specimen collected near Auke Bay, Alaska, and the lower pair from near Victoria. There is a slight difference in size and detail, but they do not appear to be significantly different. Feeding tentacles are short and clawlike and range in number from 12 to 15, with 8 to 12 side branches. When freshly dredged, swollen sections of the body look like clear jelly beans.

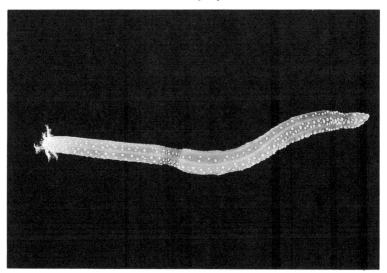

Fig. 70. Live *Chiridota albatrossii* from Saanich Inlet, B.C.

Chiridota albatrossii **Edwards, 1907 (See Photo 34)**: this species can grow up to 25 cm long. In life, it is a pinkish purple; but in alcohol the body is pale with minute orange spots. The 12 pinnate tentacles each have 6 to 14 lateral fingers. Three irregular rows of white spots occur along the dorsal side, and a few rows ventrally. Wheel ossicles measure from 80 to 120 μm in diameter. Rod-shaped ossicles in the anterior dorsal skin are 50 to 150 μm long.

Recorded from Behm Canal, southeastern Alaska, to northern Vancouver Island by Edwards (1907); Strait of Georgia (Heding 1928); in the western Pacific from Sakhalin Island south to Hokkaido (Ohshima 1915). Heding (1928) reports it from 46 to 732 m.

Chiridota discolor **Eschscholtz, 1829 (See Photo 35)**: This species can grow up to 30 cm long. It ranges in colour from whitish, yellowish, greyish, reddish to brownish: the shade depending on the abundance of red pigment in the skin and degree of contraction of the body. It has 12 tentacles, each with 4 or 5 pairs of digits. The few wheel papillae are usually confined to the dorsal side of the body in irregular rows. The diameter of the wheels is 50-105 μm. No other types of ossicles are found.

It was first described from Sitka, Alaska under stones and in loose sand. Ohshima (1915) records it from Unalaska and Umnak Island. Also recorded from Okhotsk Sea. This may be the common species found intertidally in southeastern Alaska.

Chiridota laevis **(Fabricius, 1780)**: *C. laevis* can grow up to 20 cm long, but is usually only 5 to 10 cm. It is usually pinkish, sometimes pinkish brown, transparent or colourless — rarely greyish. The wheel papillae are in 3 rows of about 20 to 30 on the dorsal side. It has 12 tentacles with 10 to 12 lateral digits. Duncan and Sladen (1881) consider *Chiridota discolor* Eschscholtz, 1829 to be synonymous, so they extend the range of *C. laevis* from the north Atlantic (Greenland, Labrador, northern Europe) to include the records of *C. discolor*.

Chiridota nanaimensis **Heding, 1928:** This species was described from six fragments collected from 46 m in Nanoose Bay, British Columbia. In alcohol it is greyish purple with bright yellow tentacles. It has 12 tentacles with 5 pairs of digits each. The wheels are 50 to 70 μm in diameter. It is supposedly differentiated from other *Chiridota* by a dense layer of microscopic crescent-shaped rods in the skin. However, *albatrossii* also has rods in the skin. The rods in the tentacles are less curved and larger. There have been no other records of this species in the literature since this original description. Whether it is a distinct species or not is unresolved.

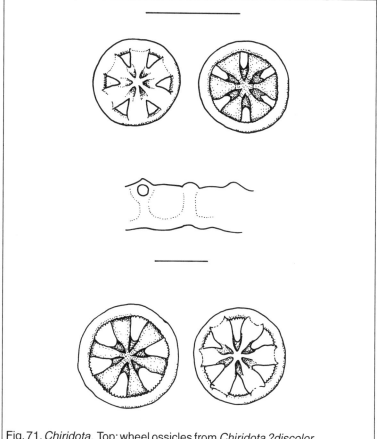

Fig. 71. *Chiridota.* Top: wheel ossicles from *Chiridota ?discolor.*
Middle: section of calcareous ring from *Chiridota ?discolor.*
Bottom: wheel ossicles from *Chiridota ?albatrossii.*
Top scale bar 100 μm; middle scale bar 1 mm.

Similar Species

Chiridota is similar to the other common wormlike genus, *Leptosynapta*, but *Leptosynapta* is usually yellow, reddish or white; it has no rows of white spots, and it inhabits muddy intertidal or shallow subtidal areas. *Leptosynapta* species have anchors and anchor plates in the skin, not wheel ossicles.

Habitat

Specimens that I have identified as *Chiridota albatrossii* are extremely abundant (55 per m²) in the soft sediment of fiords like Howe Sound, Quatsino Sound, Alice Arm, Burke Channel and Portland Canal. I have also collected it in about 16 metres (50 feet) of water near the head of Saanich Inlet, near Victoria, in a soft, shell-covered slope next to burrowing anemones (*Pachycerianthus)*. In southeastern Alaska *C. discolor* occurs at low tide under rocks in sand and gravel.

Biology

While scuba diving on the evening of 28 March 1997 in Sechelt Inlet, Neil McDaniel observed large numbers of *Chiridota* extending from a mixed shell-gravel-mud substratum and releasing gametes from the anterior end. To my knowledge, this has not been reported in the literature before.

References

Barr and Barr (1983), Brandt (1835), Clark, H. L. (1901c, 1902, 1907), Deichmann (1954), Edwards (1907), Grube (1850, 1851), Heding (1928), McDaniel et al. (1976), Ohshima (1915), Ostergren (1898), Sluiter (1901), Smirnov (1989).

ACKNOWLEDGEMENTS

I am indebted to many people over the years, from summer students who helped me collect specimens to academic colleagues who reviewed my papers on sea cucumbers. I cannot thank them all, or even remember them all individually, so I will start by thanking all those who helped out along the way but are not named here. Their contributions are greatly appreciated.

Several people provided that original spark that was so important in the beginning of my career. Dr Arthur Fontaine, from the University of Victoria, got me started on echinoderms. As an inspirational and humorous lecturer he piqued my interest in invertebrates. It was as his research assistant that I was first exposed to the microscopic anatomy of sea cucumbers and other echinoderms. The late Dr Frank Bernard, from the Pacific Biological Station in Nanaimo, facilitated the acquisition of the Station's echinoderm collection for the Royal British Columbia Museum. Dr Bernard also encouraged me to produce my first sea cucumber publication: a technical report on that collection.

Maureen Downey of the Smithsonian Institution encouraged me during the publication of my first handbook on sea stars. Dave Pawson, a sea cucumber taxonomist from the same institution, has also been inspirational to me scientifically and personally. A number of colleagues whom I first met at the International Conference of Echinoderms in Tampa have encouraged this work. Maria Byrne, Norm Sloan, Scott Smiley, Scott McEuen, John Miller, Gordon Hendler, Lane Cameron, Ailsa Clark and Frank Rowe all deserve thanks for their contributions to the field, and for personal dialogue and friendship.

I also extend my thanks to Andy Lamb, Mary Sewell, Neil McDaniel, Bill Austin, Rick Harbo, Ann Muscat, Mary Bergen, Chuck Baxter, Joe Brumbaugh for their personal field observations, specimens and other information that have been used in this book.

Many of my taxonomic studies have depended on loans of specimens from other museum collections. I extend my appreciation for loans to Peter Frank, Canadian Museum of Nature; Bob Van Syok, California Academy of Sciences; Mary Sue Brancato, Museum of Comparative Zoology; Nora Foster, University of Alaska, Fairbanks; Cindy Ahearn, Smithsonian Institution; Bruce Wing and Charles O'Clair, Auke Bay Biological Laboratory and Kathy Groves, Natural History Museum of Los Angeles County.

The financial support of the Royal British Columbia Museum and the Friends of the Royal B.C. Museum has made most of this work possible. Special thanks to Gerald Luxton of the Royal B.C. Museum's Design Section for producing the line drawings in the Introduction. I would also like to thank Alex Peden, Rob Cannings and Ted Miller, former and present supervisors, who encouraged and supported this project. In particular, I want to express my appreciation to my diving buddies, Gordon Green, Grant Hughes, Brent Cooke and a series of summer students, for some memorable times on the good ship *Nesika*. In recent years I am grateful for the diving assistance of Jim Cosgrove, Kelly Sendall and numerous other individuals.

I would like to thank Dr Rita O'Clair, Dr Frank Rowe and Dr Scott McEuen for reviewing the manuscript and providing many useful comments and suggestions for improvement.

Chris Borkent, a co-op student at the Royal British Columbia Museum at the time of publication, did a thorough job of proof-reading the text and offering helpful suggestions.

Finally, my deep appreciation goes to Tara Steigenberger who edited and produced this book, while enduring all my last minute changes and additions with a smile.

GENERAL REFERENCES

At the end of each species description I have listed sources of information and also some specialized references for that species. There are a number of popular publications that cover a wide range of marine invertebrate groups, including sea cucumbers, that are still in print or available at most libraries. I will list them here with a brief comment about each one.

The book that I turn to first, for information on intertidal biology is Ricketts, Calvin and Hedgpeth (1985), a comprehensive book that is now in its fifth edition. Ed Ricketts, one of the original authors, is a historical figure on the west coast who collected and studied the fauna from Mexico to Alaska. The book is full of facts and personal observations by this colourful individual. Books by MacGinitie and MacGinitie (1968) and Johnson and Snook (1967) are also written by biologists who were active in the first half of this century. Some of the taxonomy is outdated now, but most of the biological facts are still relevant.

More recent books such as Barr and Barr (1983), Gotshall and Laurent (1979), Kozloff (1983), Morris et al.(1980) and Snively (1978) contain photographs of some live sea cucumbers in addition to accounts of the ecology and biology of common species.

There are also several technical books on echinoderms and invertebrates. A.M. Clark (1977) and Nichols (1969) are readable texts on the biology, structure and evolution of echinoderms. Lawrence (1987) covers the functional biology of echinoderms, including nutrition, respiration, locomotion and reproduction. Hyman (1955), a standard reference for the invertebrates, has a whole volume devoted to the echinoderms of the world. Many of the illustrations in this book have appeared over and over again in biology textbooks through the years. M.F. Strathmann (1987) is an excellent reference on the reproductive biology of marine invertebrates from the north Pacific coast. I obtained much of the reproductive information for this book from McEuen (1987). Recent reviews by Smiley et al. (1991), and Harrison and Chia (1994), provide updates on sea cucumber research.

Kozloff (1987) is the only other identification key that is applicable to the waters of B.C. and Washington. The present book provides the only identification key that covers the waters from Alaska to Puget Sound. Because some species of sea cucumber look similar externally, the key requires the use of internal as well as external features, including microscopic characters such as ossicles.

REFERENCES

Alton, M.S. 1972. Bathymetric distribution of the echinoderms off the northern Oregon coast. Pp. 475-537. In *The Columbia River Estuary and Adjacent Ocean Waters,* edited by A.T. Pruter and D.L. Alverson. Seattle: University of Washington Press.

Anderson, R.S. 1965. The anatomy and burrowing behaviour of *Leptosynapta clarki.* Zoology 533 Report. Seattle: University of Washington.

------. 1966. Anal pores in *Leptosynapta clarki* (Apoda). *Canadian Journal of Zoology* 44:1031-35.

Arndt, A., C. Marquez, P. Lambert and M.J. Smith. 1996. Molecular phylogeny of eastern Pacific sea cucumbers (Echinodermata: Holothuroidea) based on mitochondrial DNA sequence. *Molecular Phylogenetics and Evolution* 6(3):425-37.

Atwood, D.G. 1973. Ultrastructure of the gonadal wall of the sea cucumber, *Leptosynapta clarki* (Echinodermata:Holothuroidea). *Zeitschrift fuer Zellforschung* 141:319-30.

------. 1974a. Fine structure of spermatogonia, spermatocysts, and spermatids of the sea cucumbers *Cucumaria lubrica* and *Leptosynapta clarki* (Echinodermata - Holothuroidea). *Canadian Journal of Zoology* 52:1389-96.

------. 1974b. Fine structure of the spermatozoon of the sea cucumber, *Leptosynapta clarki* (Echinodermata: Holothuroidea). *Cell and Tissue Research* 149:223-33.

------. 1975. Spermatogenesis and egg investments in *Leptosynapta clarki, Cucumaria lubrica* and *Cucumaria pseudocurata* (Echinodermata, Holothuroidea), with a note on acrosomal reactions. Ph.D. thesis. University of Alberta, Edmonton.

Atwood, D.G., and F.S. Chia. 1974. Fine structure of an unusual spermatozoon of a brooding sea cucumber, *Cucumaria lubrica. Canadian Journal of Zoology* 52:519-23.

Austin, W.C. 1985. *An annotated checklist of marine invertebrates in the cold temperate northeast Pacific.* Cowichan Bay, B.C.: Khoyatan Marine Laboratory.

Baker, S.M., and N.B. Terwilliger. 1993. Hemoglobin structure and function in the Rat-tailed Sea Cucumber, *Paracaudina chilensis. Biological Bulletin* 185:115-22.

Bakus, G.J. 1974. Toxicity in holothurians: A geographical pattern. *Biotropica* 6:229-36.

Barr, L. and N. Barr. 1983. *Under Alaskan seas.* Anchorage: Alaska Northwest Publishing Company.

Bergen, M. 1996. Class Holothuroidea including keys and descriptions to all continental shelf species in California. Pp. 195-250. In *Taxonomic Atlas of the benthic fauna of the Santa Maria Basin and Western Santa Barbara Channel,* Vol. 14. edited by Blake, J.A., P.H. Scott and A. Lissner. Santa Barbara Musuem of Natural History.

Berkeley, E. 1924. On a new case of commensalism between echinoderm and annelid. *Canadian Field Naturalist* 28:193.

Bingham, B.L., and L.F. Braithwaite. 1986. Defense adaptations of the Dendrochirote Holothurian *Psolus chitonoides* Clark. *Journal of Experimental Marine Biology and Ecology* 98:311-22.

Birkeland, C., P.K. Dayton and N.A. Engstrom. 1982. A stable system of predation on a holothurian by four asteroids and their top predator. *Australian Museum Memoir* 16:175-89.

Black, R.E.L. 1954. The anatomy of a parasitic gastropod in the holothurian *Parastichopus californicus.* Zoology 533 Report. Friday Harbor: University of Washington.

Boolootian, R.A. 1962. The perivisceral elements of echinoderm body fluids. *American Zoologist* 2:275-84.

Brandt, J.F. 1835. *Prodromus descriptionis animalium ab H. Mertensio in orbis terrarum circumnavigatione observatorum.* Fasciculus I: 42-62. Petropoli: Sumptibus Academiae.

Brenchley, G.A. 1982. Mechanisms of spatial competition in marine soft-bottom communities. *Journal of Experimental Marine Biology and Ecology* 60:17-33.

Brooks, E.J. 1965. Some aspects of the taxonomy and biology of the genus *Leptosynapta* (Holothuroidea) in British Columbia. M.Sc. thesis. University of Victoria.

Brumbaugh, J.H. 1965. The anatomy, diet and tentacular feeding mechanism of the Dendrochirote holothurian *Cucumaria curata,* Cowles 1907. Ph.D. thesis. Stanford University.

------. 1980. Holothuroidea: the sea cucumbers. Pp. 136-45. In *Intertidal Invertebrates of California,* edited by R.H. Morris, D.P. Abbott and E.C. Haderlie. Stanford, Ca: Stanford University Press.

Burke, R.D., D.G. Brand and B.W. Bisgrove. 1986. Structure of the nervous system of the auricularia larva of *Parastichopus californicus. Biological Bulletin* 170:450-60.

Byrne, M. 1985a. Evisceration behaviour and seasonal incidence of evisceration in the holothurian *Eupentacta quinquesemita* (Selenka). Ophelia 24:75-90.

------. 1985b. The life history of the gastropod *Thyonicola americana* Tikasingh, endoparasitic in a seasonally eviscerating holothurian host. Ophelia 24:91-101.

------. 1985c. The mechanical properties of the autotomy tissues of the holothurian *Eupentacta quinquesemita* and the effects of certain physico-chemical agents. *Journal of Experimental Biology* 117:69-86.

------. 1986. Induction of evisceration in the holothurian *Eupentacta quinquesemita* and evidence for the existence of an endogenous evisceration factor. *Journal of Experimental Biology* 120:25-39.

Cameron, J.L. 1985. Reproduction, development, processes of feeding and notes on the early life history of the commercial sea cucumber *Parastichopus californicus* (Stimpson). Ph.D. thesis. Simon Fraser University.

------. 1986. Observations on the ecology of the juvenile life stage of the California Sea Cucumber, *Parastichopus californicus* (Stimpson). *American Zoologist* 26:41.

Cameron, J.L., and P.V. Fankboner. 1984. Tentacle structure and feeding processes in life stages of the commercial sea cucumber *Parastichopus californicus* (Stimpson). *Journal of Experimental Marine Biology and Ecology* 81:193-209.

------. 1986. Reproductive biology of the commercial sea cucumber *Parastichopus californicus* (Stimpson) (Echinodermata: Holothuroidea). I. Reproductive periodicity and spawning behaviour. *Canadian Journal of Zoology* 64:168-75.

Carney, R.S., and A.G.J. Carey. 1976. Distribution pattern of holothurians on the northeastern Pacific (Oregon, U.S.A.) continental shelf slope and abyssal plain. *Thalassia Jugoslavica* 12:67-74.

Cherbonnier, G. 1951. Holothuries de l'Institut Royal des Sciences Naturelles de Belgique. *Memoires de l'Institut Royal des Sciences Naturelles de Belgique,* Series 2, no. 41:1-64.

Chia, F.S., and R.D. Burke. 1978. Echinoderm metamorphosis: fate of larval structures. Pp. 219-34. In *Settlement and metamorphosis of marine invertebrate larvae*, edited by F.S. Chia and M. Rice. New York: Elsevier.

Chia, F.S., and J.G. Spaulding. 1972. Development and juvenile growth of the sea anemone *Tealia crassicornis*. Biological Bulletin 142:206-18.

Chia, F.S., D. Atwood and B. Crawford. 1975. Comparative morphology of echinoderm sperm and possible phylogenetic implications. *American Zoologist* 15:533-65.

Clark, A.H. 1920. The echinoderms of the Canadian arctic expedition, 1913-18. *Report of the Canadian Arctic Expedition* 1913-18 8:3-13.

Clark, A.M. 1977. *Starfishes and related echinoderms.* Third Edition. Hong Kong: T.F.H. Publications Inc. Ltd (British Museum: Natural History).

Clark, A.M. and F.W.E. Rowe. 1967. Proposals for stabilization of the names of certain genera and species of holothurioidea. Z.N.(S.) 1782. *Bulletin of Zoological Nomenclature* 24:98-115.

Clark, H.L. 1901a. The holothurians of the Pacific coast of North America. *Zoologische Anzeiger* 24:162-71.

------. 1901b. Echinoderms from Puget Sound: observations made on the echinoderms collected by the parties from Columbia University, in Puget Sound in 1896 and 1897. *Proceedings of the Boston Society* 29:323-37.

------. 1901c. Synopses of North American invertebrates. XV. The holothurioidea. *American Naturalist* 35:479-96.

------. 1902. Notes on some North Pacific holothurians. *Zoologische Anzeiger* 25:562-64.

------. 1907. *The Apodous Holothurians. A Monograph of the Synaptidae and Molpadiidae.* Smithsonian Contributions to Knowledge no. 35. Washington: Smithsonian Institution.

------. 1908. Some Japanese and East Indian echinoderms. *Bulletin of the Museum of Comparative Zoology* (Harvard) 51:279-311.

------. 1913. Echinoderms from Lower California, with descriptions of new species. *Bulletin of the American Museum of Natural History* 32:185-235.

------. 1920. Reports on the scientific results of the expedition to the tropical Pacific, in charge of Alexander Agassiz, by the U.S. Fish Commission Steamer Albatross, from August 1899 to March 1900, Commander Jefferson F. Moser, USN, commanding. XXII. From October 1904 to March 1905, Commander Jefferson F. Moser, USN, commanding. XXXIII. Holothurioidea. *Memoirs of the Museum of Comparative Zoology* (Harvard) 39:121-54.

------. 1922. The holothurians of the genus Stichopus. *Bulletin of the Museum of Comparative Zoology* (Harvard) 65:39-74.

------. 1923. Echinoderms from Lower California, with descriptions of new species: supplementary report. *Bulletin of the American Museum of Natural History* 48:147-63.

------. 1924. Some holothurians from British Columbia. *Canadian Field Naturalist* 38:54-57.

Collison, J.S. 1983. *Niche differences in* Cucumaria curata *and* Cucumaria pseudocurata *of Central California.* Pacific Grove, Ca: Hopkins Marine Station.

Cowles, R.P. 1907. *Cucumaria curata* sp. nov. *Johns Hopkins University Circular* new series 3:8-9.

Da Silva, J., J.L. Cameron and P.V. Fankboner. 1986. Movement and orientation patterns in the commercial sea cucumber Parastichopus californicus (Stimpson) (Holothuroidea: Aspidochirota). *Marine Behaviour and Physiology* 12:133-47.

Deichmann, E. 1937. The Templeton Crocker expedition. IX. Holothurians from the Gulf of California, the west coast of Lower California and Clarion Island. *Zoologica (New York)* 22:161-76.

------ 1938a. Eastern Pacific expeditions of the New York Zoological Society. XVI. Holothurians from the western coasts of Lower California and Central America, and from the Galápagos Islands. *Zoologica (New York)* 23:361-87.

------. 1938b. New holothurians from the western coast of North America and some remarks on the genus Caudina. *Proceedings of the New England Zoological Club* 16:103-15.

------. 1941. The holothurioidea collected by the Velero III during the years 1932 to 1938. Part 1, Dendrochirota. *Allan Hancock Pacific Expeditions* 8:61-195.

------. 1947. Shallow water holothurians from Cabo de Hornos and adjacent waters. *Anales del Museo Argentino de Ciencias Naturales "Bernardino Rivadavia"* 42:325-51.

------. 1954. The holothurians of the Gulf of Mexico. *Fishery Bulletin of the Fish and Wildlife Service* 55:381-410.

D'jakonov, A.M. 1949. Identification of the echinoderms of the Far Eastern seas. *Bulletin of the Pacific Scientific Research Institute of Marine Fisheries and Oceanography* 30:1-132.

Duncan, P.M., and W.P. Sladen. 1881. *A memoir on the echinodermata of the Arctic Sea to west of Greenland.* London: John van Voorst.

Dybas, L., and P.V. Fankboner. 1986. Holothurian survival strategies: mechanisms for the maintenance of a bacteriostatic environment in the coelomic cavity of the sea cucumber, *Parastichopus californicus. Developmental and Comparative Immunology* 10:311-30.

Edwards, C.L. 1907. The holothurians of the North Pacific coast of North America collected by the *Albatross* in 1903. *Proceedings of the United States National Museum* 33:49-68.

------. 1910a. Four species of Pacific Ocean holothurians allied to *Cucumaria frondosa* (Gunner). *Zoologische Jahrbuecher Jena Abteilung fuer Systematik* 29:597-612.

------. 1910b. Revision of the holothuroidea. I. *Cucumaria frondosa* (Gunner) 1767. *Zoologische Jahrbuecher Jena Abteilung fuer Systematik* 29:333-58.

Ekman, S. 1923. Über *Psolus squamatus* und verwandte arten. Zugleich ein beitrag zur bipolaritätsfrage. *Arkiv foer Zoologi* 15:1-59.

------. 1925. Holothurien. *Swedish Antarctic Expedition* 1:1-194.

Ellis, D.V. 1969. Ecologically significant species in coastal marine sediments of southern British Columbia. *Syesis* 2:171-82.

Emlet, R.B. 1982. Echinoderm calcite: a mechanical analysis from larval spicules. *Biological Bulletin* 163:264-75.

Engstrom, N.A. 1974. Population dynamics and prey-predation relations of a Dendrochirote holothurian, *Cucumaria lubrica,* and sea stars in the genus *Solaster.* Ph.D. thesis. University of Washington, Seattle.

------. 1982. Brooding behaviour and reproductive biology of a subtidal Puget Sound sea cucumber, *Cucumaria lubrica* (Clark 1901)(Echinodermata: Holothuroidea). Pp. 447-50. In *Echinoderms: Proceedings of the International Conference,* Tampa Bay, edited by J.M. Lawrence. Rotterdam: A.A. Balkema.

Eschscholtz, F.R. 1829. *Atlas enhaltend abbildungen und beschreibungen neuer thierarten wahrend dis flottcapitains von kotzebue zweiter reise um die west, 1823-1826.* Berlin.

Everingham, J.W. 1961. The inter-ovarian embryology of *Leptosynapta clarki.* M.Sc. thesis. University of Washington, Seattle.

Fabricius, O. 1780. Fauna groenlandica. 8:352-57.

Fankboner, P.V. 1978. Suspension-feeding mechanisms of the Armoured Sea Cucumber *Psolus chitinoides* Clark. *Journal of Experimental Marine Biology and Ecology* 31:11-25.

Fankboner, P.V., and J.L. Cameron. 1985. Seasonal atrophy of the visceral organs in a sea cucumber. *Canadian Journal of Zoology* 63:2888-92.

Filimonova, G.F., and I.B. Tokin. 1980. Structural and functional peculiarities of the digestive system of *Cucumaria frondosa* (Echinodermata: Holothuroidea). *Marine Biology (Berlin)* 60:9-16.

Fontaine, A.R. and P. Lambert. 1973. The fine structure of the haemocyte of the holothurian, *Cucumaria miniata* (Brandt). *Canadian Journal of Zoology* 51:323-32.

------. 1976. The fine structure of the sperm of a holothurian and an ophiuroid. *Journal of Morphology* 148:209-25.

------. 1977. The fine structure of the leukocytes of the holothurian, *Cucumaria miniata*. *Canadian Journal of Zoology* 55:1530-44.

Forbes, E. 1841. *A History of British Starfishes and other Animals of the Class Echinodermata*. London: John van Voorst.

Gibson, A.W., and R.D. Burke. 1983. Gut regeneration by morphallaxis in the sea cucumber *Leptosynapta clarki* (Heding, 1928). *Canadian Journal of Zoology* 61:2720-32.

Gilliland, P.M. 1993. The skeletal morphology, systematics and evolutionary history of holothurians. *Special Papers in Palaeontology* 47:1-147.

Gotshall, D.W., and L.L. Laurent. 1979. *Pacific Coast Subtidal Marine Invertebrates: A fishwatcher's Guide*. 1st Edition. Los Osos, Ca: Sea Challengers.

Grube, A.E. 1850. Ueber die Holothurien - Gattungen Chiridota und Synapta. *Archiv fuer Anatomie, Physiologie, und wissenschaftliche Medicin* 111-17.

------. 1851. Ueber *Chiridota discolor* Eschsch. Pp. 11-18. In *Echinodermen*, edited by F. Brandt and E. Grube.

Gunnerus, J.E. 1767. *Beskrifning pa trenne Norska Sjo-krak, Sjopungar Kallade*. Pp. 114-24. In *Kongelige Vet. Akademie Handlingar.* Stockholm.

Hadfield, M.G. 1961. The morphology of a psolid holothurian. M.Sc. thesis. University of Washington, Seattle.

Hansen, B., and J.D. McKenzie. 1991. A taxonomic review of northern Atlantic species of Thyonidiinae and Semperiellinae (Echinodermata: Holothuroidea: Dendrochirotida). *Zoological Journal of the Linnean Society* (1986) 103:101-28.

Harrison, F.W. and F.S. Chia (eds). Microscopic anatomy of invertebrates. Echinodermata. Vol. 14. Toronto: Wiley-Liss.

Heding, S.G. 1928. Papers from Dr Th. Mortensen's Pacific Expedition, 1914-16. XLVI. Synaptidae. *Dansk Naturhistorisk Forening i Kobenhaven Videnskabelige Meddelelser* 85:105-324.

------. 1931. Uber die Synaptiden des Zoologischen Museums zu Hamburg. *Zoologische Jahrbuecher Jena Abteilung fuer Systematik* 61:637-98.

------. 1942. Holothuroidea Part II - Aspidochirota - Elasipoda - Dendrochirota. *The Danish Ingolf Expedition*, 4(13):1-39.

Heding, S.G., and A. Panning. 1954. Phyllophoridae: Eine bearbeitung der polytentaculaten Dendrochiroten holothurien des zoologischen museums in Kopenhagen. *Spolia Zoologica Musei Hauniensis* 13:1-209.

Hess, H., B. Bingham, S. Cohen, R.K. Grosberg, W. Jefferson and L. Walters. 1988a. Population genetics of *Leptosynapta clarki*, a hermaphroditic holothuroid with limited dispersal p. 799 (abstract). In *Echinoderm Biology. Proceedings of the Sixth International Echinoderm Conference, Victoria, 1987*, edited by R.D. Burke, P.V. Mladenov, P. Lambert and R.L. Parsley. Rotterdam: A.A. Balkema.

------. 1988b. The scale of genetic differentiation in *Leptosynapta clarki* Heding, an infaunal brooding holothuroid. *Journal of Experimental Marine Biology and Ecology* 122:187-94.

Hetzel, H.R. 1960. Studies on Holothurian Coelomocytes with special reference to *Cucumaria miniata*. Ph.D. thesis. University of Washington, Seattle.

Hozawa, S. 1928. On the changes occuring with advancing age in the calcareous deposits of *Caudina chilensis* (J. Muller). *Scientific Reports of Tohoku Imperial University, Series 4* 3:361-78.

Hufty, H.M. 1973. Studies on the possible occurrence of an endogenous hormone in the ovary of *Parastichopus californicus*. M.Sc. thesis. Washington State University, Pullman.

Hufty, H.M., and P.C. Schroeder. 1974. A hormonally active substance produced by the ovary of the holothurian *Parastichopus californicus*. *General and Comparative Endocrinology* 23:348-51.

Hyman, L.H. 1955. *The Invertebrates: Echinodermata – The Coelomate Bilateria*, Vol. 4. Toronto: McGraw-Hill.

Imaoka, T. 1980. Observation on *Psolus squamatus* (Koren) from the Okhotsk Sea (Dendrochirota: Psolidae). *Publications of the Seto Marine Biological Laboratory* 25:361-72.

Inaba, D. 1930. Notes on the development of a holothurian, *Caudina chilensis* (J. Müller). *Scientific Reports of Tohoku Imperial University, Series 4* 5:215-48.

Jespersen, A., and L. Lützen. 1971. On the ecology of the Aspidochirote sea cucumber *Stichopus tremulus* (Gunnerus). *Norwegian Journal of Zoology* 19:117-32.

Johnson, M.E., and H.J. Snook. 1967. *Seashore Animals of the Pacific Coast.* Paperback Edition. New York: Dover Publishing Inc.

Johnson, M.W., and L.T. Johnson. 1950. Early life history and larval development of some Puget Sound echinoderms: with special reference to *Cucumaria* spp. and *Dendraster excentricus*. In *Studies Honouring Trevor Kincaid*, edited by M.H. Hatch. Seattle: University of Washington Press.

Jones, M. 1963. Responses of *Parastichopus californicus* (Stimpson) to asteroids. Zool. 533 Report, University of Washington, Friday Harbor.

Jones, S.A. 1962. Early embryology of *Psolus chitonoides* Clark. M.Sc. thesis. University of Washington, Seattle.

Jordan, A.J. 1972. On the ecology and behavior of *Cucumaria frondosa* (Echinodermata: Holothuroidea) at Lamoine Beach, Maine. Ph.D. thesis. University of Maine at Orono.

Kawamoto, N. 1927. The anatomy of *Paracaudina chilensis* (J. Muller) with special reference to the perivisceral cavity, the blood and the water-vascular systems in their relation to the blood circulation. *Scientific Reports of Tohoku Imperial University, Series 4* 2:239-64.

Kincaid, T. 1964. A gastropod parasitic on the holothurian, *Parastichopus californicus* (Stimpson). *Transactions of the American Microscopical Society* 83:373-76.

Kirkendale, L., and P. Lambert. 1995. *Cucumaria pallida*, a new species of sea cucumber from the northeast Pacific Ocean (Echinodermata, Holothuroidea). *Canadian Journal of Zoology* 73:542-51.

Klugh, A.B. 1923. The habits of *Cucumaria frondosa*. *Canadian Field Naturalist* 37:76-78.

Koren, J. 1844. Beskrivelser over *Thyone fusus og Cuvieria squamata. Nyt. Mag. f. Naturvid.* 4:203-225, plates i-iii.

Kozloff, E.N. 1983. *Seashore Life of the Northern Pacific Coast.* Seattle: University of Washington Press.

------. 1987. *Marine invertebrates of the Pacific Northwest.* Seattle and London: University of Washington Press.

Krishnan, S., and T. Dale. 1975. Ultrastructural studies on the testes of *Cucumaria frondosa* (Holothuroidea: Echinodermata). *Norwegian Journal of Zoology* 23:1-15, 32 figs.

Lambert, P. 1984. British Columbia marine faunistic survey report: holothurians from the northeast Pacific. *Canadian Technical Report of Fisheries and Aquatic Sciences* 1234:1-30.

------. 1985. Geographic variation of calcareous ossicles and the identification of three species of eastern Pacific sea cucumbers (Echinodermata: Holothuroidea). Pp. 437-443. In *Echinodermata: Proceedings of the Fifth International Echinoderm Conference, Galway,* edited by B.F. Keegan and B.D.S. O'Connor. Rotterdam: A.A. Balkema.

------. 1986. Northeast Pacific holothurians of the genus *Parastichopus* with a description of a new species *Parastichopus leukothele* (Echinodermata). *Canadian Journal of Zoology* 64:2266-72.

------. 1987. A new kind of cucumber. *Discovery: News and events from the Royal British Columbia Museum* 15:5.

------. 1990a. Nature's Designs. *Royal British Columbia Museum Notes.*

------. 1990b. A new combination and synonymy for two subspecies of *Cucumaria fisheri* Wells (Echinodermata: Holothuroidea). *Proceedings of the Biological Society of Washington* 103:913-21.

------. 1996. *Psolidium bidiscum,* a new shallow-water psolid sea cucumber (Echinodermata: Holothuroidea) from the northeastern Pacific, previously misidentified as *Psolidium bullatum* Ohshima. *Canadian Journal of Zoology* 74:20-31.

------. In press. A taxonomic review of five northeastern Pacific sea cucumbers (Holothuroidea). In *Proceedings of the Ninth International Echinoderm*

Conference, San Francisco, edited by R. Mooi and M. Telford. Rotterdam: A. A. Balkema.

Lawrence, J. 1987. *A functional biology of echinoderms.* London and Sydney: Croom Helm.

Layton, J.L. 1975. The distribution of *Leptosynapta clarki* at two mudflats on San Juan Island and some behavioral observations. Zool. 533 Report, University of Washington, Friday Harbor.

Leighton, B.J. 1988. A record of an echinoderm host of *Melanella columbiana* (Gastropoda: Eulimidae). *The Veliger* 31:135.

Levin, V. 1984. *Duasmodactyla kurilensis* (Dendrochirota, Phyllophoridae) -- a new species of dendrochirote holothurian from Onekotan Island (the Kuril Islands). *Biologiya Morya* 4: 69-72.

Levin, V.S., V.I. Kalinin, S.N. Fedorov and S. Smiley. 1986. The structure of triterpene glycosides and the systematic position of two holothurians of the family Stichopodidae. *Mar. Biol. (Vladivostok)* 1986: 72-77.

Lowenstam, H.A. and G.R. Rossman. 1975. Amorphous, hydrous, ferric phosphatic dermal granules in Molpadia(Holothuroidea): physical and chemical characterization and ecologic implications of the bioinorganic fraction. *Chemical Geology* 15:15-51.

Ludwig, H. 1875. Beiträge zur Kenntnis der Holothurien. *Arbeiten aus dem Zoolog.-zootom. Institut in Würzburg* 2:77-120 (1-42).

------. 1881. Revision der Mertens-Brandt'schen holothurien. *Zeitschrift fur Wissenschaftliche Zoologie* 35:575-99.

------. 1886. Echinodermen des Beringsmeeres. *Zoologiscen Jahrbuchern. Zeitschrift fur Systematik, Geographie und Biologie der Thiere* 1886:275-96.

------. 1894. Reports on an exploration off the west coasts of Mexico, Central and South America, and off the Galapagos Islands, in charge of Alexander Agassiz, by the U.S. Fish Commission Steamer *Albatross* during 1891: 12. The Holothurioidea. *Memoirs of the Museum of Comparative Zoology at Harvard College* 17:1-183, plates. 1-19.

------. 1900. Arktische und subarktische Holothurien. I. Family Holothuriidae (Aspidochirotae). *Fauna Arctica* 1:136-78.

Lütken, C.F. 1857. *Oversigt over Grönlands echinodermata.* Kjöbenhavn.

Lutzen, J. 1979. Studies on the life history of *Enteroxenos* Bonnevie, a gastropod endoparasitic in aspidochirote holothurians. *Ophelia* 18:1-51.

MacGinitie, G.E., and N. MacGinitie. 1968. *Natural History of Marine Animals.* 2nd edition. Toronto: McGraw-Hill.

Madsen, F.J. 1953. Holothurioidea. *Report of the Swedish Deep-sea Expedition 1947-1948* 2:149-73.

Manwell, C. 1959. Oxygen equilibrium of Cucumaria miniata hemoglobin and the absence of the Bohr effect. *Journal of Cellular and Comparative Physiology* 53:75-83.

Margolin, A.S. 1976. Swimming of the sea cucumber *Parastichopus californicus* (Stimpson) in response to sea stars. *Ophelia* 15:105-14.

Mauzey, K.P., C. Birkeland and P.K. Dayton. 1968. Feeding behaviour of asteroids and escape responses of their prey in the Puget Sound region. *Ecology* 49:603-19.

McDaniel, N.G. 1973. A survey of the benthic macroinvertebrate fauna and solid pollutants in Howe Sound. Fisheries Research Board of Canada Technical Report 385.

McDaniel, N.G., R.D. MacDonald, J.J. Dobrocky and C.D. Levings. 1976. Biological surveys using in-water photography at three ocean disposal sites in the Strait of Georgia, British Columbia. Canadian Fisheries Marine Service Technical Report 713

McDermid, M.A. 1983. Aspects of the reproductive biology of the viviparous, apodous holothurian *Leptosynapta clarki* (Heding). B.Sc. thesis. University of Victoria.

McEdward, L.R., and F.S. Chia. 1991. Size and energy content of eggs from echinoderms with pelagic lecithotrophic development. *Journal of Experimental Marine Biology and Ecology* 147:95-102.

McEuen, F.S. 1986. The reproductive biology and development of twelve species of holothuroids from the San Juan Islands, Washington. Ph.D. thesis. University of Alberta.

------. 1987. Phylum Echinodermata, Class Holothuroidea. Pp. 574-96. In *Reproduction and development of marine invertebrates of the northern Pacific coast,* edited by M.F. Strathmann. Seattle: University of Washington Press.

------. 1988. Spawning behaviours of northeast Pacific sea cucumbers (Holothuroidea: Echinodermata). *Marine Biology (Berlin)* 98:565-85.

McEuen, F.S. and F.S. Chia. 1985. Larval development of a molpadiid holothuroid, *Molpadia intermedia* (Ludwig 1894) (Echinodermata). *Canadian Journal of Zoology* 63:2553-59.

------. 1991. Development and metamorphosis of two psolid sea cucumbers, *Psolus chitonoides* and *Psolidium bullatum,* with a review of reproductive patterns in the family Psolidae (Holothuroidea: Echinodermata). *Marine Biology (Berlin)* 109:267-79.

McKenzie, J.D. 1991. The taxonomy and natural history of north European dendrochirote holothurians (Echinodermata). *Journal of Natural History* 25:123-71.

Miller, J.E., and D.L. Pawson. 1984. Holothurians (Echinodermata: Holothuroidea). *Memoirs of the Hourglass Cruises,* 7:1-79.

Mitsukuri, K. 1912. *Studies on Actinopodous Holothurioidea.* Tokyo: University of Tokyo.

Morris, R., ed. 1995. *British Columbia Sea Cucumber Newsletter.* 1(1):1-4.

Morris, R.H., D.P. Abbott and E.C. Haderlie. 1980. *Intertidal Invertebrates of California.* Stanford, Ca: Stanford University Press.

Mortensen, T. 1927. *Handbook of the Echinoderms of the British Isles.* Reprint 1977, Rotterdam: Dr W. Backhuys, Uitgever.

------. 1932. The Godthaab Expedition 1928 – Echinoderms. *Meddelelser om gronland* 79:3-62.

Mottet, M.G. 1976. The fishery biology and market preparation of sea cucumbers. Washington Department of Fisheries Technical Report, no. 22.

Müller, J. 1850. *Anatomische Studien über die Echinodermen. Archiv fuer Anatomie und Physiologie* 1850:117-55, 225-33.

Nichols, D. 1969. *Echinoderms.* London: Hutchinson University Library.

Nichols, F.H. 1975. Dynamics and energetics of three deposit-feeding benthic invertebrate populations in Puget Sound, Washington. *Ecological Monographs* 45:57-82.

Nomura, S. 1926. Influence of oxygen tension on the rate of oxygen consumption in Caudina. *Scientific Reports of Tohoku Imperial University, Series 4* Biology:133-38.

Nordhausen, R.W. 1972. Tentacle ultrastructure and production of feeding mucus in the dendrochirote holothurian *Cucumaria pseudocurata* Deichmann 1938. M.Sc. thesis. California State College, Sonoma.

O'Foighil, D., and A. Gibson. 1984. The morphology, reproduction and ecology of the commensal bivalve *Scintillona bellerophon* spec. nov. (Galeommatacea). *The Veliger* 27:72-80.

Ohshima, H. 1915. Report on the holothurians collected by the U.S. fisheries steamer *Albatross* in the N.W. Pacific during the summer of 1906. *Proceedings of the United States National Museum* 48:213-91, plates. 8-11.

Ostergren, H. 1896. Zur kenntnis der subfamilie Synallactinae unter den Aspidochiroten. Pp. 348-60. *Festskrift für Lilljeborg.* Upsala.

Ostergren, H. 1898. Das system der Synaptiden. *Öfversigt af Kongl. Vetenskaps-Akadamiens Forhandlingar* 2:111-20.

Ostergren, H. 1902. The holothurioidea of northern Norway. *Bergens Museums Arbok* 9:1-32.

Panning, A. 1955. Bemerkungen Uber die Holothurien-Familie Cucumariidae (Ordnung Dendrochirota). *Mitteilungen aus dem Hamburgischen Zoologischen Museum und Institut* 53:33-47.

------. 1962. Bemerkungen über die holothurien-familie Cucumariidae (Ordnung Dendrochirota) 3. Teil. Die gattung Pseudocnus Panning 1949. *Mitteilungen aus dem Hamburgischen Zoologischen Museum und Institut* 60:57-80.

Pawson, D.L. 1963. The holothurian fauna of Cook Strait, New Zealand. Zoology Publications from Victoria University of Wellington, no. 36.

------. 1965. The bathyal holothurians of the New Zealand region. Zoology Publications from Victoria University of Wellington, no. 39.

------. 1970. The marine fauna of New Zealand: sea cucumbers (Echinodermata: Holothuroidea). Bulletin of the New Zealand Department of Scientific and Industrial Research, no. 201.

------. 1977. Marine flora and fauna of the northeastern United States. Echinodermata: Holothuroidea. NOAA Technical Report Circular, no. 405.

------. 1982. Holothuroidea. Pp. 813-18. In *Synopses and classification of living organisms,* edited by S.P. Parker. Toronto: McGraw-Hill.

Pearse, J.S., D.J. McClary, M.A. Sewell, W.C. Austin, A. Perez-Ruzafa and M. Byrne. 1988. Simultaneous spawning of six species of Echinoderms in Barkley Sound, British Columbia. *Invertebrate Reproduction and Development* 14:279-88.

Perrier, R. 1902. Holothuries. *Expedition Scientifique du "Travailleur" et du "Talisman",* 5:273-554.

------. 1905. Holothuries Antarctiques du Museum d'Histoire Naturelle de Paris. *Annales des Sciences Naturelles Zoologie,* series 9, 1-2.

Ricketts, E.F., J. Calvin and J.W. Hedgpeth. 1985. *Between Pacific Tides.* 5th Ed. Stanford, Ca: Stanford University Press.

Runnstrom, J., and S. Runnstrom. 1921. Uber die Entwicklung von *Cucumaria frondosa* Gunnerus und *Psolus phantapus* Strussenfelt. *Bergens Museums Arbok,* no. 5, plates I-VIII.

Rutherford, J.C. 1973. Reproduction, growth and mortality of the holothurian *Cucumaria pseudocurata. Marine Biology (Berlin)* 22:167-76.

Sabourin, T.D. and W.B. Stickle. 1981. Effects of salinity on respiration and nitrogen excretion in two species of echinoderms. *Marine Biology (Berlin)* 65:91-99.

Selenka, E. 1867. Beiträge zur anatomie und systematik der holothurien. *Zeitschrift fur Wissenschaftliche Zoologie* 17:291-374.

Semper, C. 1868. *Reisen im Archipel der Phillippinen. Zweiter Theil Wissenschlaftliche Resultate Erster Band I: Holothurien.* Wiesbaden.

Sewell, M.A. 1994a. Birth, recruitment and juvenile growth in the intraovarian brooding sea cucumber Leptosynapta clarki. *Marine Ecology-Progress Series* 114:149-56.

------. 1994b. Small size, brooding, and protandry in the apodid sea cucumber *Leptosynapta clarki. Biological Bulletin* 187:112-23.

Sewell, M.A., and F.S. Chia. 1994. Reproduction of the intraovarian brooding apodid Leptosynapta clarki (Echinodermata: Holothuroidea) in British Columbia. *Marine Biology (Berlin)* 121:285-300.

Sewell, M.A., A.S. Thandar and F.S. Chia. 1995. A redescription of *Leptosynapta clarki* Heding (Echinodermata: Holothuroidea) from the northeast Pacific, with notes on changes in spicule form and size with age. *Canadian Journal of Zoology* 73:469-85.

Shick, M.J. 1983. Respiratory gas exchange in echinoderms. Pp. 67-110. In *Echinoderm Studies,* vol. 1, edited by M. Jangoux and J.M. Lawrence. Rotterdam: A.A. Balkema.

Shimek, R.L. 1987. Sex among the sessile. *Natural History* 96:60-63.

Shinn, G.L. 1983. *Anoplodium hymenae* sp. n., an umagillid turbellarian from the coelom of *Stichopus californicus,* a Northeast Pacific holothurian. *Canadian Journal of Zoology* 61:750-60.

------. 1986. Life history and function of the secondary uterus of *Wahlia pulchella,* an umgallid turbellarian from the intestine of a northeastern Pacific sea cucumber *(Stichopus californicus). Ophelia* 25:59-74.

Sloan, N.A. 1984. Echinoderm fisheries of the world: a review. Pp. 109-24. In *Proceedings of the Fifth International Echinoderm Conference, Galway, Ireland, 24-29 September 1984*, edited by B.F. Keegan and B.D.S. O'Connor. Rotterdam and Boston: A.A. Balkema.

------. 1986a. Underwater World – Sea Cucumber. Communications Directorate, Department of Fisheries and Oceans, Ottawa. (pamphlet)

------. 1986b. World jellyfish and tunicate fisheries, and the northeast Pacific echinoderm fishery. Pp. 23-33. In *North Pacific Workshop on Stock Assessment and Management of Invertebrates*, Canadian Special Publication of Fisheries and Aquatic Sciences 92, edited by G.S. Jamieson and N. Bourne. Ottawa: Fisheries and Oceans Canada.

Sluiter, C.P. 1901. *Die holothurien der Siboga-Expedition. Siboga-Expedition*, no. 44.

Smiley, S. 1984. A description and analysis of the structure and dynamics of the ovary, of ovulation, and of oocytes maturation in the sea cucumber *Stichopus californicus*. M.Sc. thesis, University of Washington, Seattle.

------. 1986a. Metamorphosis of *Stichopus californicus* (Echinodermata: Holothuroidea) and its phylogenetic implications. *Biological Bulletin* 171: 611-31.

------. 1986b. *Stichopus californicus:* oocyte maturation hormone, metamorphosis, and phylogenetic relationships. Ph.D. thesis, University of Washington.

------. 1988a. The dynamics of oogenesis and the annual ovarian cycle of *Stichopus californicus* (Echinodermata: Holothuroidea). *Biological Bulletin.* 175: 79-93.

------. 1988b. Investigation into purification and identification of the oocyte maturation hormone of *Stichopus californicus* (Holothuroidea: Echinodermata). Pp. 541-49. In *Echinoderm Biology*, edited by R.D. Burke, P.V. Mladenov, P. Lambert and R.L. Parsley. Rotterdam: A.A. Balkema.

------. 1994. Holothuroidea. Pp. 401-71. In *Microscopic Anatomy of Invertebrates*, vol. 14, edited by F.W. Harrison and F.S. Chia. Toronto: Wiley-Liss.

Smiley, S., and R.A. Cloney. 1985. Ovulation and the fine structure of the *Stichopus californicus* fecund ovarian tubule. *Biological Bulletin.* 169: 142-364.

Smiley, S., F.S. McEuen, C. Chaffee and S. Krishnan. 1991. Echinodermata: Holothuroidea. Pp. 663-750. In *Reproduction of Marine Invertebrates*, vol. 6, edited by A.C. Giese, J.S. Pearse and V.B. Pearse. Pacific Grove, Ca: Boxwood Press.

Smirnov, A.V. 1989. A new species of holothurian *Trochodota inexspectata* (Synaptida: Chiridotidae) from the Simushir Island (Kuril Islands). *Zoologicheskii Zhurnal* 68:155-60.

Smith, E.H. 1962. Studies of *Cucumaria curata* Cowles 1907. *Pacific Naturalist* 3:233-46.

Smith, N.S. 1966. Activity and movement of *Parastichopus californicus* in relation to density and water depth. Zoology 533 Report. University of Washington, Friday Harbor.

Smith, R.I., and J T. Carlton. 1975. *Light's Manual. Intertidal Invertebrates of the Central California Coast.* Berkeley, Ca: University of California Press.

Snively, G. 1978. *Exploring the Seashore in British Columbia, Washington and Oregon: A Guide to Shorebirds and Intertidal Plants and Animals.* West Vancouver: Gordon Soules Book Publishers.

Stickle, W.B., and C.J. Denoux. 1976. Effects of in situ tidal salinity fluctuations on osmotic and ionic composition of body fluids in S.E. Alaska rocky intertidal fauna. *Marine Biology (Berlin)* 37:125-35.

Stimpson, W. 1857. Crustacea and Echinodermata of the Pacific shores of North America. *Boston Journal of Natural History* 6:524-31.

Stimpson, W. 1864. Descriptions of new species of marine invertebrata from Puget Sound, collected by the naturalists of the North-west Boundary Commission, A.H. Campbell, Esq., Commissioner. *Proceedings of the Academy of Natural Sciences of Philadelphia* 16:153-61.

Strathmann, M.F. 1987. *Reproduction and Development of Marine Invertebrates of the Northern Pacific Coast: Data and Methods for the Study of Eggs, Embryos, and Larvae.* Seattle and London: University of Washington Press.

Strathmann, R.R. 1978. Length of pelagic period in echinoderms with feeding larvae from the Northeast Pacific. *Journal of Experimental Marine Biology and Ecology* 34:23-27.

Strathmann, R.R. and H. Sato. 1969. Increased germinal vesicle breakdown in oocytes of the sea cucumber *Parastichopus californicus* induced by starfish radial nerve extract. *Experimental Cell Research* 54:127-29.

Stricker, S.A. 1985. The ultrastructure and formation of the calcareous ossicles in the body wall of the sea cucumber *Leptosynapta clarki* (Echinodermata, Holothuroida). *Zoomorphology (Berlin)* 105:209-22.

Stricker, S.A. 1986. The fine structure and development of calcified skeletal elements in the body wall of holothurian echinoderms. *Journal of Morphology* 188:273-88.

Sutterlin, A.M., and S. Waddy. 1976. Tentacle movement patterns involved in feeding behaviour of the sea cucumber *Cucumaria frondosa. Marine Behaviour and Physiology* 4:17-24.

Swan, E.F. 1961. Seasonal evisceration in the sea cucumber, *Parastichopus californicus* (Stimpson). *Science* 133:1078-79.

Taghon, G.L. 1982. Optimal foraging by deposit feeding invertebrates: Roles of particle size and organic coating. *Oecologia* 52:295-304.

Tao, L. 1930. Notes on the ecology and physiology of *Caudina chilensis* (Müller) in Matsu Bay. *Proceedings of the Fourth Pacific Science Congress* 3(1929):7-11.

Terwilliger, R.C. 1975. Oxygen equilibruim and subunit aggregation of a holothurian hemoglobin. *Biochimica et Biophysica Acta* 386:62-68.

Terwilliger, R.C., and K.R.H. Read. 1970. The hemoglobins of the holothurian echinoderms *Cucumaria miniata* Brandt, *Cucumaria piperata* Simpson [sic], and *Molpadia intermedia* Ludwig. *Comparative Biochemical Physiology and Comparative Physiology* 36:339-51.

Théel, H. 1886. Report on the Holothurioidea dredged by the HMS *Challenger* during the years 1873-1876 Part II. *Report of the Scientific Results of the Voyage of H.M.S.* Challenger *1873-76,* Zoology 14:1-290.

Thompson, W. 1840. Contributions towards a knowledge of the Mollusca Nudibranchia and Mollusca Tunicata of Ireland, with descriptions of some apparently new species of Invertebrata. *Annals of Natural History* 5:84-102.

Tikasingh, E.S. 1960. Endoparasitic gastropods of some Puget Sound holothurians. *Journal of Parasitology (Supplement)* 46:13.

------. 1961. A new genus and two new species of endoparasitic gastropods from Puget Sound, Washington. *Journal of Parasitology* 47:268-72.

Turner, R.L., and J.C. Rutherford. 1976. Organic, inorganic, and caloric composition of eggs, pentaculae, and adults of the brooding sea cucumber *Cucumaria curata* Cowles (Echinodermata: Holothuroidea). *Journal of Experimental Marine Biology and Ecology* 24:49-60.

von Britten, M. 1906. Holothurien aus dem Japanischen und Ochotskischen Meere. *Bulletin de l'Academie Imperiale des Sciences de St. Petersbourg* 25:123-57.

Wells, H. 1924. New species of Cucumaria from Monterey Bay, California. *Annals and Magazine of Natural History, Series 9* 14:113-21.

Wolcott, T.G. 1981. Inhaling without ribs: the problem of suction in soft bodied invertebrates. *Biological Bulletin* 160:189-97.

Yamanouchi, T. 1926. Some preliminary notes on the behaviour of the holothurian, *Caudina chilensis* (J. Müller). *Scientific Reports of Tohoku Imperial University, Series 4* 2:85-91.

Young, C.M. and F.S. Chia. 1982. Factors controlling spatial distribution of the sea cucumber *Psolus chitonoides:* settling and post-settling behaviour. *Marine Biology (Berlin)* 69:195-205.

GLOSSARY

Ampulla: a bulblike expansion at the base of a tube foot that contracts or expands to extend or retract the foot (plural: ampullae).

Anaesthetized: being in a relaxed state and insensitive to pain or feeling.

Auricularia: a stage in the larval development of a sea cucumber (Fig. 8E).

Calcareous: containing calcium carbonate, a form of limestone or chalk.

Chiton: a class of molluscs characterized by eight overlapping dorsal plates.

Cloaca: in sea cucumbers, the intestine and the respiratory trees empty their products into this chamber before passing to the outside.

Coelom: the main body cavity that contains the intestine, gonad and respiratory trees.

Commensal: a close association between two organisms in which one derives an advantage while the other neither gains nor loses.

Cosmopolitan: describes a species having a worldwide distribution wherever a suitable habitat occurs.

Cuvierian organs or tubules: a cluster of white sticky tubes attached to the cloaca of some tropical sea cucumbers that are extruded to entangle a predator.

Dendritic: refers to the treelike branching pattern of a tentacle (Fig. 2).

DFO: Department of Fisheries and Oceans, Canada.

Digitate: a type of short stubby tentacle with a few side branches (Fig. 2).

DNA: Deoxyribonucleic acid, building blocks of the genetic material in cells.

Doliolaria: old name used for a late larval stage of a sea cucumber; now referred to as a pelagic juvenile.

Dorsal: refering to the dorsum, the back or upper surface of an animal.

Echinoderm: a group (phylum) of spiny-skinned marine animals that includes sea stars, sea urchins, sea lilies, sea cucumbers, brittle stars and sea daisies.

Ethanol: a type of alcohol used as a preservative.

Evisceration: in sea cucumbers, a spontaneous expulsion of the internal organs through a split in the body wall, or via the cloaca.

Formalin: a 40% solution of formaldehyde gas in water, used as a fixative or disinfectant.

Gastropod: a class of molluscs including snails and slugs.

Gonad: reproductive organ that produces eggs or sperm.

Gonopore: opening through which the eggs or sperm pass to the outside.

Holotype: a single specimen chosen by the author to represent a new species and is deposited in a museum collection.

Introvert: thin-walled anterior end of some sea cucumbers that can be drawn inside the body along with the tentacles.

IQ: Individual Quota; the maximum weight of a species that can be harvested in a season.

Lecithotrophic: refers to a larva that does not actively feed, but lives off a store of yolk.

Madreporite: in sea cucumbers, an internal calcareous nob at the end of a tube that communicates between the coelomic fluid and the water-vascular system.

Magnesium chloride: a chemical used to anaesthetize marine animals prior to preservation.

Magnesium sulfate: Epsom salts; also used as an anaesthetic for aquatic animals.

Menthol crystals: anaesthetic used for aquatic animals.

Mounting medium: a liquid resin used to mount thin objects between a cover slip and glass slide for viewing with a microscope.

Negatively phototactic: moves away from light source.

Nematode: a phylum of unsegmented worms often called roundworms; includes a number of parasitic worms.

Ossicles: tiny crystals of calcium carbonate (calcite) found in the skin of sea cucumbers; each species has ossicles of a unique shape (Fig. 5).

Papillae: any small extension of the body wall; the California Sea Cucumber has many pointed papillae on its back.

Parasitic: describes an organism that lives on or in another and causes some harm to it.

Paratype: specimens other than a holotype that are designated to document the variability of a new species.

Peltate: a type of feeding tentacle in sea cucumbers that resembles a mop (Fig. 2).

Pentaradial symmetry: a shape resembling a five-spoked wheel in which each of the five segments is identical.

Peristaltic: refers to waves of contraction that move along a tube, such as the intestine or the body of a sea cucumber.

Phosphatic: refers to a salt or ester of phosphoric acid.

Pinnate: a type of feeding tentacle with a main stem and numerous single side branches (Fig. 2).

Planktonic: refers to animals or plants (plankton) that drift with the current.

Planktotrophic: a form of planktonic larva that feeds on other plankton.

Podia: tube feet of echinoderms.

Polian vesicles: blind sacs that attach to the oral ring of the water-vascular system of sea cucumbers; thought to act as a fluid reservoir (Fig. 3).

Polychaete: a class of marine segmented worms.

Protandric hermaphrodite: an animal that produces both eggs and sperm, but at different times.

Respiratory trees: treelike organs situated in the body cavity and connected to the cloaca. They draw water in and transfer oxygen to the internal fluids of the sea cucumber (Fig. 3).

Rete mirable: a knot of finely branching blood vessels between two loops of the intestine.

Sole: flat underside of a sea cucumber of the Family Psolidae.

Submersible: small manned submarine.

Substratum: the ground or object that an organism lives on or in (plural: substrata).

Subterminal: a position on the underside of a body near the posterior or anterior end.

Suspension feeding: any feeding method that utilizes particles floating in the water.

Taxonomist: a scientist who studies the naming and classification of animals or plants.

Taxonomy: the study of naming and classifying organisms.

Type specimens: specimens that form the basis of a species description; the holotype is a single specimen chosen by the author to represent a new species and is deposited in a museum collection; paratypes are other specimens designated to document the variability of a species.

Ventral: the underside of an organism closest to the ground; opposite of dorsal.

Vitellaria larva: early larval stage of a direct developing sea cucumber (Fig. 8).

Water-vascular system: a closed network of tubes in echinoderms that operates the tube feet and tentacles like a hydraulic system.

INDEX

CREDITS

The cover images, *Psolus chitonoides* (front) and *Parastichopus californicus* (back), were photographed by James A. Cosgrove of the Royal British Columbia Museum.

Illustration by F.S. McEuen on page 17 © by University of Washington Press, used with permission.

Figures in Body © RBCM
All photos in black and white by Philip Lambert except for:
Brent Cooke, RBCM: Figures: 11, 60
James A. Cosgrove, RBCM: Figure: 9

Colour Photo Insert © RBCM
Brent Cooke, RBCM: 2, 5, 9, 10, 11, 15, 29, 36, 37
James A. Cosgrove, RBCM: 1, 34
Phil Lambert, RBCM: 3, 4, 6, 7, 12, 13, 14, 16-28, 30-33, 35
C.E. O'Clair: 8

Sea Cucumbers of British Columbia, Southeast Alaska and Puget Sound
Edited, designed and produced by Tara Steigenberger, RBCM.
Typeset in Times New Roman 11/14. Main titles in Helvetica.
Cover Design by Chris Tyrrell, RBCM.
Illustrated by Gerald Luxton and Philip Lambert, RBCM.
Proofread by Chris Borkent.
Printed and bound in Canada by Friesens.